The Real World of Victorian Steampunk

The Real World of Victorian Steampunk

Steam Planes and Radiophones

Simon Webb

PEN & SWORD **HISTORY**

AN IMPRINT OF PEN & SWORD BOOKS LTD.
YORKSHIRE – PHILADELPHIA

First published in Great Britain in 2019 by
Pen & Sword History
An imprint of
Pen & Sword Books Ltd
Yorkshire - Philadelphia

Copyright © Simon Webb, 2019

ISBN 978 1 52673 285 9

Printed and bound in England
By TJ International Ltd.

Pen & Sword Books Ltd incorporates the Imprints of Pen & Sword Books
Archaeology, Atlas, Aviation, Battleground, Discovery, Family History, History,
Maritime, Military, Naval, Politics, Railways, Select, Transport, True Crime,
Fiction, Frontline Books, Leo Cooper, Praetorian Press, Seaforth Publishing,
Wharncliffe and White Owl.

For a complete list of Pen & Sword titles please contact

PEN & SWORD BOOKS LIMITED
47 Church Street, Barnsley, South Yorkshire, S70 2AS, England
E-mail: enquiries@pen-and-sword.co.uk
Website: www.pen-and-sword.co.uk

or

PEN AND SWORD BOOKS
1950 Lawrence Rd, Havertown, PA 19083, USA
E-mail: uspen-and-sword@casematepublishers.com
Website: www.penandswordbooks.com

Contents

List of Plates

Introduction

In January 2018 a company chaired by British businessman Sir Richard Branson proposed building passenger-carrying tubes to link London's three airports, Heathrow, Gatwick and Stanstead. It would, it was announced, take just five minutes to get from Heathrow to Gatwick by this means, the carriages travelling at an estimated 670mph; approximately the speed of sound. This was a scaled-down version of a scheme proposed the previous year by American entrepreneur Elon Musk, pioneer of privately-funded space travel, who claimed to have been given permission to build a tunnel stretching 200 miles, from New York to Washington DC. Trains would be propelled at 700mph along this so-called 'Hyperloop' by creating a vacuum and so assuring that there would be no air pressure to slow down the train. Newspaper reports in July 2017 described the scheme as being 'futuristic' and 'space-age'. Accompanying the press release were artists' impressions of what this new railway might look like, were it to be built. The whole enterprise sounded very up-to-date and modern; an exciting innovation in travel for the twenty-first century.

There was though something curiously familiar about these supposedly new ideas for high-speed travel. Illustration 1 shows a drawing published in 1829. This seems to be an early version of something which looks uncannily like the Hyperloop, labelled on the side, 'Grand Vacuum Tube Company, Direct to Bengal'. In other words, rather than running for a mere 200 miles between two cities, this was a proposed intercontinental vacuum railway, one which would carry passengers from Britain all the way to the region comprising present-day Bangladesh, a distance of 5,000 miles or so. It must be said that this casts Elon Musk's suggested rapid transit system into the shade. In 1895, almost 70 years after the drawing of the intercontinental vacuum tube transit system appeared, *The Strand Magazine* in London published a fictional account of a transatlantic tunnel, linking Liverpool and the American city of Boston. Passengers would travel along this vacuum tube at over 1,000 miles per hour, crossing the Atlantic in just two and three-quarter hours.

Surely though, all this is just Victorian fantasy? There could not have been any real possibility of a vacuum railway in the nineteenth century, could there? Illustration 2 in fact shows this very thing in operation; a vacuum railway tube operating in London's Crystal Palace Park in 1867. It ran for just 600 yards, but those prepared to part with 6d, a little under 3p in decimal currency, could be drawn along a tunnel in the same way that the New York-to-Washington Hyperloop might one day operate. Nor was this the first railway system to exploit vacuums and air pressure to draw trains along without their being pulled by locomotives. As early as 1847, trains working on this principle were running between the English towns of Exeter and Newton Abbot. The journey time of 20 minutes was faster than the electric trains running on this route today. And in 1870, an atmospheric or pneumatic transport system began operating in a tube beneath New York.

Both the Crystal Palace, South Devon and New York pneumatic railways were powered by stationary steam engines, which meant that passengers were not troubled with all the smoke, steam, dirt and noise which were inescapable features of travel by steam trains at that time. The nineteenth century was, after all, the age of steam and it was used for every conceivable purpose; powering printing presses and factories, as well as transport on land and at sea. Vacuum railways sometimes feature in novels and short stories belonging to a genre of science fiction called 'steampunk'. Harry Harrison's *A Transatlantic Tunnel, Hurrah!*, first published in 1972, tells the story of a vacuum railway being constructed in a tube which lies on the seabed of the Atlantic Ocean and will connect Britain and America.

The word 'steampunk' may require a little explanation. The *Oxford English Dictionary* says that 'steampunk' is 'a genre of science fiction that typically features steam-powered machinery rather than advanced technology'. This is a very rough-and-ready definition and steampunk enthusiasts often have their own, more idiosyncratic ideas of what constitutes true steampunk. For some purists, the narrative must be set in the nineteenth century and preferably resemble the world of H.G. Wells or Jules Verne. No sooner had the OED included 'steampunk' in their dictionary, than complaints were heard that the word 'Victorian' should have been used in the definition. There are those though who reject the need for 'steampunk' and 'Victorian' to be inextricably linked. They are inclined to include in the steampunk canon books set in the modern day, which depict an alternative world which has evolved following some point of divergence in history; the Confederates winning the American Civil War, or the failure of the American Revolution, for example. Such worlds are, almost invariably, more backward than our own, at least from a purely technological perspective. In the present book, we shall adopt the most catholic, all-embracing and elastic definition of steampunk, drawing upon the widest possible variety of sources.

The first recorded use of the word 'steampunk' dates back to 1987 and 30 years later, steampunk has expanded to include not only literature, but also computer games, graphic novels, art, clothing, various accessories such as walking sticks, fob-watches and glasses, and even music. The word refers now not merely to a type of science fiction, but to an entire style of fashion.

Although the expression 'steampunk' was coined only a little over 30 years ago, the concept itself has been around for a good deal longer than that. Early instances of science fiction or fantasy tales which feature anachronistic technology, steam-powered aeroplanes say, are sometimes referred to as being 'proto-steampunk'. *Queen Victoria's Bomb*, by Ronald Clark, was published in 1968 and is often cited as being a 'proto-steampunk' work. The plot concerns the development and testing of a nuclear bomb in the nineteenth century and its proposed use to bring a decisive end to the Crimean War fought between 1853 and 1856. *Bring the Jubilee*, by Ward Moore, was first published in 1953 and is another book which is reckoned by many to belong to the category of 'proto-steampunk'. Set in 1950s America, the petrol engine has not been developed and steam-powered 'minibiles' are the only self-powered road vehicles. Heavier-than-air flight is also unknown and steam engines drive the balloons which sail overhead.

In *A Transatlantic Tunnel, Hurrah!*, set in an alternative 1973, there are aeroplanes, but there is a neat twist. Instead of being powered by kerosene, as in our world, the fuel which they use is pulverized coal dust. While the tunnel of the title is being created, these coal-fired aeroplanes fly overhead, belching smoke into the sky. Steam aircraft are commonly to be found in the world of steampunk. This can surely be only a preposterous fantasy? There may have been vacuum trains in the real world, but it can hardly be the case that steam planes were flying across the skies of Victorian England? We turn though to the 21 September 1894 edition of that respected American journal, *Scientific American*, and read, to our surprise;

> On Tuesday, July 31, for the first time in the history of the world, a flying machine actually left the ground, fully equipped with engines, boiler, fuel, water, and a crew of three persons.

Nine years before the first flight of the Wright brothers and we are reading an account of a steam-powered aeroplane taking to the air in the English county of Kent. This was, by the way, no flimsy construction of wood and canvas, such as the Wright brothers were later to experiment with. The aeroplane described in the *Scientific American* was made of steel, weighed three-and-a-half tons and had a wingspan of 125ft. It may be seen in Illustration 13. By way of comparison, a well-known modern airliner, the Airbus 320, has a wingspan of only 117ft. In Illustration 3 we can see a drawing from half a century earlier than the aircraft described in the *Scientific American*. This fantastic contraption, which might very easily be supposed to have escaped from the pages of a steampunk novel, was intended to be the backbone of the world's first international airline.

In this book, we shall be looking at an unfamiliar world of steam planes and radiophones, at a mechanical Internet and cable news service, as well as the first self-propelled buses in the world, which were steam-powered and ran regular services in London and other British cities in the 1820s and 1830s, to say nothing of a steam-powered racing car which reached the almost unbelievable speed of 127mph. It is the world often portrayed in steampunk fantasies, except that this world is real. The fact is that although steampunk is a specialized fantasy genre, it is actually founded in historical events, most of which have been forgotten. Who today remembers the first wireless telephone, which was invented in the 1870s? This forerunner of today's mobile phones can be seen in Illustration 4. Fears of exhausting the supply of fossil fuels to be found in the earth are nothing new either. Solar power was being exploited many years before the death of Queen Victoria. Illustration 9 shows us a solar power plant. The parabolic mirrors in both this and the radiophone seen in Illustration 4 look strangely modern and both pictures look like classic examples of steampunk art, where modern inventions appear in unfamiliar guises in Victorian settings.

The fact is that the real Victorian world so frequently resembled that depicted and referenced in modern steampunk, that it is often hard to distinguish between the two. Imagine, for a moment, electric jewellery; brooches, tie-pins and hair clips which are battery-powered, so that they light up and move. A miniature skull which can be worn attached to a shirt and whose eyes flash, while the jaw clatters up and down. It is precisely the sort of novelty which one might find today if you were to go on Amazon and search for 'steampunk'.

In 1885, electric jewellery, including such items as glowing owls to pin one's shawl and an illuminated electric walking stick, were to be found in the brochure produced by Gustave Trouve in Paris. The battery was kept in the pocket or handbag and when required, could be connected up, so that a butterfly in a woman's hair could light up and flap its wings, for instance. One feels sure that there must be a market for mechanical insects and skulls of this sort among today's aficionados of steampunk.

We will begin our exploration of the real world of steampunk by examining the origins of modern steampunk, which lie both in forgotten Victorian inventions and also the speculative fiction which was popular in the nineteenth century. These, combined with the so-called proto-steampunk novels of the 1960s and 1970s, set the scene for the steampunk genre as we know it today and allow us to see clearly how it developed in the way that it has.

Chapter 1

Dreams of the Future,
Visions of the Past

The enchantment of steampunk fiction lies in the juxtaposition of either weirdly modern inventions being used by men and women from the Victorian Era or old-fashioned items such as airships and semaphore towers still being used in the twentieth or twenty-first century. The first of these may be called 'classic' steampunk; a perfect example being William Gibson and Bruce Sterling's seminal novel *The Difference Engine*. The computers and cars in a High Victorian setting found in the book are oddly unsettling and create a fantastic world similar to, but distinctly different from, that which we know from history books. Little wonder that *The Difference Engine* is seen as a perfect example of Victorian steampunk literature. Keith Roberts' *Pavane*, first published in 1968, is of the second type of steampunk, where the modern world is restricted to technologically-inferior methods of communication and travel; semaphore towers are still the fastest form of communication in the early twenty-first century and steam vehicles rule the roads. Most novels of this second kind are alternative universes, where some point of divergence in the past from the history with which we are familiar has led the development of a strangely altered version of the world we actually live in. In *Pavane*, this point was the assassination of Elizabeth I in 1588 and the subsequent invasion of Britain by Spanish forces, securing the worldwide power of the Catholic Church for the next 400 years.

Images from what may be called 'classic steampunk', that which is set during the nineteenth century and is reminiscent of H.G. Wells or Jules Verne, are indistinguishable at times from those which actually date from that period. For many aficionados of steampunk, the closer that a work resembles the world of Verne or Wells, the better and more authentic is it likely to be. For some enthusiasts, the word 'Victorian' should invariably precede the word 'steampunk'. For them, alternative present worlds which are dependent upon nineteenth-century technology do not really count as steampunk.

Although the expression 'science fiction' itself dates only from the twentieth century, the concept is much older than that. In 1666 Margaret

Cavendish wrote *The Blazing World*, which has been described by some as the first science fiction story. It tells of an invasion of part of the earth by the inhabitants of another planet, who use submarines and aerial warfare to subdue their enemies on earth. It was in the mid-nineteenth century though that this form of literature really came into its own and it is there that the roots of steampunk are to be found. Writers such as Jules Verne created worlds peopled by Victorian gentlemen like Captain Nemo, who gallivanted around in futuristic submarines. Some flew to the moon in spaceships which were astonishingly similar to the one which actually took American astronauts round the moon in the 1960s.

Just as writers of steampunk today look back to the Victorian era and refashion it by combining strange and sometimes anachronistic technology with the clothes and styles of that time, so too did some nineteenth-century authors try to look forward and imagine a future filled with wonders which were unknown to their own time. The results of such speculative fiction are uncannily similar to some of today's steampunk, in that we see Victorian men and women engaging with spaceships, robots, aeroplanes and death rays, while at the same time operating steam-powered or clockwork machinery. Both types of fiction are based upon little-known incidents and inventions of the real world. Sometimes the connections between real events in the world, speculative fiction of the late nineteenth century which references this and also a modern steampunk story, are so close that the joins cannot be seen. A classic case of this is Boilerplate, a supposedly Victorian robot which has been described as 'the gateway drug to steampunk'.

In 2000 husband and wife team Paul Guinan and Anina Bennett had it in mind to create a steampunk graphic novel featuring a robot. Instead, they made a website and began to create doctored photographs of real people and scenes from the late nineteenth and early twentieth centuries, inserting the image of their 12in-high model robot, Boilerplate. They played this with completely straight faces, crafting a back-story for Boilerplate which involves an inventor called Archibald Campion, who first displayed his robot in 1893 at the World's Columbian Exposition in Chicago.

The inspiration for Boilerplate came from American dime novels, published between 1868 and 1900; some of which feature metal men who are eerily similar in appearance to Boilerplate. *The Steam Man of the Prairies*, written by Edward S. Ellis and published in 1868, told the story of a steam-powered metal robot and the adventures which he had with his young creator. Other authors picked up on the idea of the metal men and there was a succession of cheap novels published which had as plot devices robots powered by steam or electricity. These provide us with some of the earliest examples of American science fiction. None of them, of course, actually mentioned 'robots', this word only being coined in 1920.

It is plain that Paul Guinan and his wife drew extensively upon the images and plots of the nineteenth-century fiction which had as its protagonists metal men. They freely adapted original illustrations from such books and incorporated them first into their website and later into fictitious histories of Victorian robots, which they then published in book form. Guinan estimated that approximately a third of the visitors to his website found the material there so convincing that they left believing Boilerplate to have been a real invention.

So far, we see a modern author of steampunk cannibalizing Victorian fiction to enhance his own fictitious work, but both the dime novels and Guinan's own writing and illustrations are connected with genuine inventions. In *The Steam Man of the Prairies*, a steam-powered, humanoid robot runs along, pulling behind him a carriage. This may be seen in Illustration 6. There are unmistakable similarities between this image and Boilerplate. However, both have a common origin in the real world.

A report in the *New York Express* of 21 March 1868 includes an account of an invention which was on display in New York and without doubt formed the basis for firstly the steam man in Edward Ellis' story and also Boilerplate, the twenty-first century steampunk robot. After saying a few words about Zadoc Dederick, the inventor of the Steam Man, the article continues as follows;

Mr. Steam Man is a person of commanding presence, standing 7ft 9in in his stocking vamps, weighs 500 pounds, measures 200 inches round the waist, and decidedly bucolic in general appearance. At this early hour in the morning he was rather in dishabille, and minus his pants. This circumstance, though detracting rather from his comeliness, was yet more than counterbalanced by the greater facilities it gave for the study of human anatomy, and was eagerly availed of for that purpose. The legs are made of iron cranks, screws, springs ad infinitum, not quite as attractive in exterior as those we see in the weekly pictorials, but evidently of greater durability and strength. The motion of the legs is almost facsimile to that of the human extremities, and the manner in which they are set agoing strikingly calls to mind the philosophic apostrophe of the human donkey to his namesake, 'How fearfully and wonderfully we are made.' The abdominal region is occupied by a good-sized furnace, which was in full blast. The steam man's boiler is delicately concealed from the profanity of the public gaze, but is presumed to be somewhere above the furnace. This complex piece of machinery once got out of order, but was happily restored after a careful investigation of the cause and the application of the appropriate remedy. The steam whistle is fixed in his mouth, the gauge at the back of the head, and the safety valve in an appropriate position. He wears a large stove-pipe hat — stove-pipe literally, for it is through the cranium the funnel passes. His hands are gloved, a good moustache ornaments his face, and in outward garb he is rather good-looking than otherwise.

The photograph which accompanied this piece showed a stout metal figure, dressed like man, who was attached to the front of a cart and apparently ready to pull it along. According to the inventor, this metal man could not only pull a wagon along at 60 miles per hour, he was also quite capable of stepping over obstructions. It has to be said that despite the fact that the inventor was promising to market these remarkable machines at just $300 each, he and his robot soon vanished from sight. It seems highly likely that the whole thing was a confidence trick of some kind.

There can be little doubt that the steam man was the prototype in the first instance for the eponymous 'Steam Man of the Prairies' and then, by extension for Boilerplate. The cover of *The Steam Man of the Prairies* shows a steam-powered metal man pulling a cart, just as in Zadoc Dederick's invention. Since the newspaper reports appeared in March 1868 and the dime novel in August of that same year, it is not difficult to see who copied whom. We thus see a perfect instance of how nineteenth-century speculative fiction, combined with a genuine invention from the same period, have mingled together and been transmuted into present-day steampunk.

Another example of the process by which steampunk has evolved should make this even clearer. This is the idea of a railway tunnel across the Atlantic Ocean, along which trains travelling through a vacuum will move at tremendous speed. At first sight, this is an absurd notion, but it also happens to be one with its roots in both historical fact and also Victorian fiction. Just like Boilerplate, the two have been ingeniously combined to produce a steampunk novel.

In 1865 construction began in London of a pneumatic or air-powered railway which would ferry passengers under the River Thames, from

Waterloo Station on the south bank to Whitehall, propelling the carriages by air pressure alone. An Act of Parliament had been passed which authorized the building of this line, whose trains would travel along huge iron tubes laid on the riverbed. A banking crisis caused the collapse of the company undertaking the construction of this tube link, but for a while there was great excitement about the possibilities of such a mode of travel, especially since it was to pass under water. It was suggested that a similar link might be constructed, in the same way, between England and France. Extremely fast trains would then travel across the Channel as regularly as rail services to Manchester or Birmingham.

The idea of a metal tube carrying passengers under the water at great speed was an intriguing one and it appeared as the central theme of a story published 30 years later. *The Strand Magazine* came out every month and was well known in Victorian Britain for the fiction which appeared in it. It was in *The Strand Magazine* that Sherlock Holmes made his debut and it was in the November 1895 issue that a short story by Michel Verne, son of the more famous Jules, was published. This was called *An Express of the Future* and described a journey through a 3,000 mile-long vacuum tube laid across the Atlantic. Like Richard Branson and Elon Muske's recent proposals, trains travelled through the tunnel which Michel Verne described at astonishing speed, something over 1,000mph.

The idea of an intercontinental vacuum tunnel had of course been around for almost 70 years when Michel Verne used it as the basis for his story in *The Strand*. We turn again to Illustration 1 to see that this idea had already been mooted, albeit in satirical form, as early as 1829. Following the thread which leads from fiction to fact, back to fiction and then to fact, culminating in the projects which people like Richard Branson are now proposing, is not always easy.

So far, so good. We have seen a genuine project in Victorian Britain used as the inspiration for a science fiction-type story by the son of Jules Verne. How does this tie in directly with the modern genre of steampunk? Harry Harrison's *A Transatlantic Tunnel, Hurrah!*, which first appeared in 1972, is often cited as a good example of the proto-steampunk genre. It describes an alternative universe where the Americans lost the War of Independence and Britain continues to rule the entire North American continent. In this world, in 1973, it is decided that a railway should be built across the Atlantic Ocean, to connect Britain and America. The coal-burning aeroplanes which are the fastest form of transport in this alternative universe are not really sophisticated enough to link the various parts of the world effectively.

The railway which is discussed in *A Transatlantic Tunnel, Hurrah!* is to run through a sealed tube which will be laid across the floor of the Atlantic Ocean. This is a massive engineering project, which is discussed in some detail. The technical details are almost identical to those described in the story by Michel Verne. In fact, *An Express of the Future* and *A Transatlantic Tunnel, Hurrah!* are astonishingly similar in various ways. One of these similarities is a common theme in modern steampunk and is worth remarking upon here.

In *An Express of the Future*, the characters talk and behave like Victorians. This is scarcely surprising, since it was written and published when Victoria was queen. However, the characters in *A Transatlantic Tunnel, Hurrah!* also talk and act like Victorians, although the narrative is actually set in 1973. Travel in London is still by Hansom cabs, although these are now electrical rather than horse-drawn, and the conversations between driver and passenger sound as though they could be taking place in the nineteenth rather than the twentieth century. The protagonist of the novel, Captain Augustus Washington, is engaged to a young woman, with whose father he falls out. As a consequence, his fiancé's father forbids him the house and unilaterally breaks off the engagement between the two young people. One might readily imagine such high-handed action being taken by a Victorian father, but hardly by one in 1973, the year in which the story is set. This kind of approach, where a story set in the modern world features people who speak as though they have escaped from a book by Jules Verne or H.G. Wells is quite a common conceit in steampunk novels. In other words, the strange dissonance, which is the very essence of steampunk, when a character is seen to be engaging with modern technology, is also extended to the manners and general behaviour of the people in the story. This odd sense of modern technology combined with Victorian *mores* and behaviour is also a notable feature of the early science fiction novels which were so popular from the 1860s onwards. In *Le Vingtième siècle; La vie électrique* (*The Twentieth Century; The Electric Life*) by Albert Robida, for example, published in 1890, we have such scenes as a father discussing his daughter's dowry; an exceedingly nineteenth-century topic of conversation. The twist is that the conversation takes place via Skyping or something as similar. The telephonoscope enables each party to see the other during a conversation which takes place at some distance. Although he was perfectly capable of visualizing a future filled with various marvellous inventions, Robida was unable to imagine any corresponding social changes.

Writing in the nineteenth century, Robida shows us his idea of the modern world, with many aspects of information and communication technology which are recognizable, but the people behave in quaint and outdated ways. This is similar to the effect of steampunk alternative universes. Keith Roberts' *Pavane* was published in 1968 and set in the late twentieth and early twenty-first centuries, but the feudal system has survived and Latin and Middle English are still spoken in England. In *Bring the Jubilee*, we see an alternative version of the United States in the 1940s, in a world where the Confederates won the American Civil War. The protagonist attends a political rally in New York for the Whig party. The atmosphere of *Bring the Jubilee* very much suggests the late nineteenth century, rather than the 1940s. One final example is Michael Moorcock's *The Warlord of the Air*, set in the 1970s but redolent of the height of the British Empire at about the time of the Edwardians.

The genres of modern steampunk and nineteenth-century speculative fiction blur into each other, until it is impossible to tell one from the other. An illustration for one of Robida's novels shows a family watching a flat-screen television which is bringing them live coverage of a war in a distant country. The hairstyle and clothes, combined with the décor of the home, tell us at once that we are firmly in the Victorian age. The television relaying the news however throws us a little, because it does not accord with what we would expect to see at this time. We have the same feeling when looking at Illustration 18, which shows a spaceship splashing down on the Pacific Ocean, after having flown around the moon. This is a very curious image indeed, because it is associated with some of the most accurate forecasting of scientific developments imaginable. It illustrates a novel in which Victorian gentlemen fly to the moon.

Victorian spaceships are precisely the sort of thing which we might expect to come across in a steampunk narrative. The one shown in Illustration 18 is a remarkable example of such a thing, because it is not the product of some modern writer who has attempted to transpose the American Apollo programme to the mid-nineteenth century, but was rather produced by a man living a century or more before the events which he described.

The first mission to the moon, which did not land on the surface but merely orbited the moon and returned to earth to splash down in the ocean, took place in 1968. Years before this flight, there was a stiff competition between the American states of Texas and Florida to see which of them would have the glory of hosting the launching pad for missions into orbit and to the moon. Florida won the prize of having the base where the spaceships took off from, Cape Canaveral, but Texas was awarded the honour of hosting the base for mission control, which was to be in Houston.

In 1969 the spaceship *Columbia*, with a crew of three astronauts, took off from the Florida coast and headed for the moon. It took four days and six hours to reach the moon, travelling at a speed of 24,000mph. A hundred years earlier, a spaceship was launched from a giant gun called *Columbiad*, also from the Florida coast. It too had a crew of three and it took four days and one hour to reach the moon, travelling at 25,000mph. Both *Columbia* and the spaceship fired from the *Columbiad* splashed down in the Pacific Ocean on returning to Earth and the crews of both spaceships were picked up by naval vessels.

The *Columbiad* was of course a wholly fictional creation of Jules Verne's and appears in *From the Earth to the Moon*, published in 1865. It is set in what was then the modern day. It is books such as this which originally inspired the whole idea of steampunk. The plainly Victorian men and women who may be seen in the illustrations to Jules Verne's books are pictured with all manner of advanced technological gadgetry and could have stepped straight out of a modern steampunk novel. A little-known work of Jules Verne's, which although written in the 1860s was not published until 1994, it provides us with a good example of the second kind of steampunk story; that set in a version of the modern world which is eerily different from the one which we know.

Modern steampunk novels of the alternative universe type show a strangely altered and distorted modern world, with unfamiliar technology and people who often behave as though they more properly belonged in the nineteenth century. A novel written in 1863, called *Paris in the Twentieth Century*, could easily pass as belonging to this genre. When Jules Verne submitted it for publication, there were not yet such things as internal combustion engines, telephones, gramophones or tube trains, yet all appear in the book, which portrays life in France in the year 1960. There is even a version of the Internet; a global communication network, accessed via electro-mechanical terminals. Magnetically-powered trains also make an appearance, traveling at great speed along sealed tubes. This is perhaps the earliest incarnation in written fiction of the 'Hyperloop' schemes suggested in recent years by Elon Musk and Richard Branson.

The fascinating things about *Paris in the Twentieth Century*, and the aspect which makes it so similar to certain modern steampunk novels, is that although the action takes place in the mid-twentieth century, it is peopled by men and women who talk and act like Victorians, even down to their preoccupation with the dichotomy between art and science. In 1863, there was a great deal of debate about whether science and technology would smother literature and creative art, rendering them irrelevant to most ordinary people. We see echoes of this controversy in the works of William Morris and Charles Dickens; it is a central theme, for example, in *Hard Times*, published in 1854. Jules Verne wrote his futuristic fantasy at the height of this ideological struggle. Although by 1960, the year in which *Paris in the Twentieth Century* is set, this was no longer anything which concerned people, Verne's characters, and indeed the entire plot of the novel, involves little else.

The novels mentioned above are all either proto-steampunk, books written before the expression 'steampunk' had been coined, or speculative fiction written when Victoria was still on the throne. In *The Difference Engine*, written by William Gibson and Bruce Sterling, first published in 1990 and widely regarded as the first real steampunk novel, we have exactly the same atmosphere as is to be found in the earlier works. It is a world of computer programmers, racing-car drivers and khaki-clad British soldiers who are armed with automatic rifles. The twist is that this is all during the 1850s and that regardless of the startling developments in various fields, the characters conduct themselves and speak as one would expect Victorian Londoners of that time to do. This is fair enough; in a book set in the middle of the nineteenth century, the language used by those operating astonishingly modern devices should be typical of the time. What is interesting though is that precisely the same thing occurs in alternative universes, versions of the modern world where some point of divergence has caused the world to develop along different, and usually more backward, lines.

We saw earlier how Harry Harrison's *A Transatlantic Tunnel, Hurrah!* probably had its origins in a short story published towards the end of Queen Victoria's reign. An unkind person might even say that the 1972 novel had copied its central premise from Michel Verne's story. Looking a little closer at Harry Harrison's book reveals something curious; that it may well contain the germ of the idea which formed the basis for *The Difference Engine*. Here is a case of a real-life event inspiring one steampunk or proto-steampunk novel, which in turn led to the development of another novel; one which kicked off the modern steampunk movement.

The Difference Engine shows an alternative universe, one where mechanical, steam-driven computers bring about an information revolution in Victorian Britain. The idea for this novel comes of course from the calculating machines designed and partially built by Charles Babbage. We shall be looking at these in a later chapter. In *A Transatlantic Tunnel, Hurrah!*, set in 1973, Charles Babbage's difference engines also feature. Just as in the original machines planned by Babbage, the computers in this book consist of spinning metal shafts and cogwheels; they are entirely mechanical. However, unlike those in *The Difference Engine*, the driving shafts are turned not by steam but by electricity. Bizarrely, the spelling of Babbage's name has been altered slightly, so that we read of a 'Brabbage engine'. Here is a description of the thing in operation;

> He opened a door with a flourish to disclose serried ranks of slowly turning silver discs, all of them perforated with large numbers of small holes. Metal fingers riding on rods brushed the surfaces of the discs, bobbing and clicking when they encountered the openings.

This is a calculating machine with brass camshafts which need to be oiled regularly. It is in fact a modified version of a Babbage engine.

There is no reason to suppose that Bruce Sterling and William Gibson copied the idea of working Babbage engines from Harry Harrison; it is more that both writers engage in the same looting of the past to provide the material for their imaginary worlds. This looting consists, as we have seen in the case of both the robots and the underwater railway tunnel, in part of taking obscure inventions from the past and changing them subtly, and also by lifting anything useful from other stories, both those published in the Victorian period and more recent works.

The visual images of steampunk work in just this way, by taking whatever might be needed from the historical past and then tinkering with it. This is how Boilerplate was created, by looking at the illustration from *The Steam Man of the Prairies*, examining patent applications from the 1860s and then altering and adapting them, before melding everything together into a believable whole. The pictures illustrating a lot of speculative fiction from the late nineteenth century could very easily be placed in a modern steampunk novel and would fit in seamlessly with the genre. Look at Illustration 19, which shows the interior of Jules Verne's spaceship to the moon. This looks as comfortable as can be, like a Victorian gentleman's study, with even a dog relaxing on the furniture. This is a perfect example of both modern steampunk style and also nineteenth-century speculative fiction at its most daring and imaginative.

Stories such as that of the launching of a spaceship from a gun called *Columbiad*, which travels around the moon before splashing down in the Pacific are intriguing because they might not have merely foretold the future, but actually helped to shape it. This could be less a case of art imitating life, in the form of a book, and more life imitating art. This will require a little explanation.

Many of the ingenious devices upon which we rely today, especially those relating to transport and telecommunications, had their origins in the Victorian period. It was a time of frantic innovation and creativity, with men such as Thomas Edison, Hiram Maxim, Clement Ader and Valdemar Poulsen, of whom we will be learning more later in this book, turning out many of the gadgets which we today take for granted. Among these were record players, telephone answering machines, tape recorders, electric light, mains electricity, aeroplanes and silencers for car engines, to name but a few. These men were of course a product of their own age, they did not work and invent in a vacuum.

Some inventions and innovations which emerged in the late nineteenth century had been around as ideas for many years. Sometimes, things such as heavier-than-air flying machines, to give one example, were discussed in learned journals. More often, they entered the public consciousness via the fiction of writers like Jules Verne. Some people, of course, read articles called 'On Aerial Navigation', published in *The Journal of Natural Philosophy, Chemistry and the Arts* decades before an aeroplane existed in real life. Many, many more read *The Clipper of the Clouds* by Jules Verne. Similarly, there were those who kept up to date, via scientific publications, with the work of Paul Gottlieb Nipkow and the implications of this for what was described as 'seeing by electricity'. Those people were vastly outnumbered though by the readers of *Punch* magazine who, in 1878, saw a cartoon by Daphne du Maurier's grandfather, which showed parents Skyping, on a wall-mounted screen, with their daughter in Sri Lanka. This may be seen in Illustration 10.

In short, the public knew, through fictional works, both literature and cartoons, that things such as television, aeroplanes and even spaceships might one day be possible. Inventors working at that time hoped to make their fortunes by providing people with what they wanted and did not yet have. If speculative fiction had caused people to consider the possibility of television or aeroplanes and persuaded them that such devices would be desirable, then there was an incentive for men such as Edison and Ader to set to work to make these dreams come true.

It is this idea, of the link between the Victorian forerunners of science fiction and the inventions which were sometimes inspired by the books, short stories and visual images produced by these men, which provides the theme for this book. Modern steampunk is inextricably linked to both nineteenth-century speculative fiction and also some of the more obscure technology which was to be found at that time, such as the optical telegraphs and steam-powered aircraft. At times, the link between the fiction of the nineteenth century and the real world of the twentieth and twenty-first is easy to understand. Jules Verne described a marvellous submarine called the *Nautilus* in his 1870 book, *Twenty Thousand Leagues Under the Sea*. It was hardly surprising that this name should be chosen for the first American nuclear submarine the following century. In the same way, Verne's use of the name for his spaceship-launching gun, the *Columbiad*, referencing as it did the famous explorer Christopher Columbus, was also a natural choice for the Americans in 1969, when they wanted a name for their own spaceship which was going to the moon.

One way and another, today's steampunk has been fashioned by the fictional narratives of the nineteenth century, combined with the inventions which were being made at that time. We have looked at one or two instances of this, such as steam-powered robots. In the next chapter, we shall examine in detail two transport systems which used steam power; both of which feature in twentieth-century steampunk; namely steam-powered cars and railway trains which travel very quickly through metal tubes from which some of the air has been evacuated.

Chapter 2

Of Steam Buses and
Atmospheric Railways

A recurring feature of steampunk stories is the idea of steam-powered road vehicles. Today, we tend to assume that steam was used in Victorian times to drive railway locomotives and paddle steamers, or perhaps stationary steam engines that operated machinery. We certainly don't think of it as being a means of propulsion on the open road. In steampunk fiction though, cars, lorries and buses are often seen to be reliant on steam, the internal combustion engine never having been invented or not exploited for road transport. One example of this was mentioned in the introduction, the minibiles of 1950s America in *Bring the Jubilee*; there are many others. Keith Roberts' 1968 novel, *Pavane*, a 'fix-up' consisting of a number of previously-published short stories put together in a single volume, is a vision of England as it might have been if Elizabeth I had been assassinated in 1588. Set in the late twentieth century, *Pavane* shows a land without railways or motor cars, where the most effective means of transport are steam lorries. These 'steamers' haul trains of trailers along the roads. In *The Warlord of the Air*, a proto-steampunk novel by Michael Moorcock, steam cars are shown to be common in the alternative world of 1973.

Perhaps the most interesting appearance of steam-driven road vehicles is in *The Difference Engine*, regarded by some as the first, true steampunk novel. In it, we are introduced to 'steam-gurneys'. These are cars and buses which are steam-driven. So efficient are these 'gurneys', that some are streamlined and take part in races. The novel is set in 1855 and it is made plain that gurneys have been around for some considerable time, long enough to be refined and modified into various different types of vehicle.

The gurneys which feature in *The Difference Engine* provide a good jumping-off point for our exploration of the real world of steampunk in Victorian Britain. The reason for this is that they are not a figment of William Gibson and Bruce Sterling's imagination, but actually existed and, for a while at least, were thought to be the future of transport in nineteenth-century Britain. In the years before Victoria ascended the throne, steam-powered road vehicles produced by Gurney were seen on the streets of London and even provided a regular, scheduled service between the West Country cities of Cheltenham and Gloucester.

Goldsworthy Gurney, after whom the gurneys of *The Difference Engine* were named, was something of a Renaissance Man. An archetypal gentleman-scientist of the early nineteenth century, Gurney was a doctor, architect, chemist and inventor. Born in the Cornish town of Padstow in 1793, Gurney moved to London in 1820, by which time he had already been running his own medical practice for seven years. During this time, he had also experimented with various laboratory gadgets, including an oxy-hydrogen blowpipe. For this, he was awarded a gold medal by the Royal Society in 1823. Already though, he was working upon an invention which would, he hoped, make his name and revolutionize transport, first in Britain and then across the world.

These were exciting times for steam power. In 1825, the same year that Gurney applied for a patent for 'An apparatus for propelling carriages on common roads or railways', the Stockton and Darlington Railway opened. Although to begin with the locomotives of the new railway drew only goods wagons loaded with coal, it was not long before passenger coaches too were being pulled along the tracks by steam. The first regular passenger service by steam trains began running in 1830, between Canterbury and Whitstable. The future was looking bright for railways, as investors clamoured to put their money into new lines and the British government made a £100,000 exchequer loan to boost the industry. Strangely though, the first steam engines to carry passengers in Britain were operating not on railways, but along ordinary roads.

After patenting his ingenious new vehicles, Gurney set up a factory near Regent's Park and also started a company to promote his machines. The Gurney Steam Carriage Company was founded in 1826. It didn't take long for the young man to realize that he was onto something. At first, he built a stagecoach with a steam engine at the rear. The boiler was positioned beneath the passenger seats and this made some people uneasy. This steam carriage drove from the Regent's Park workshop to Hampstead and Highgate and then even further, to the village of Barnet, a few miles outside London. It was fast, travelling at about 20 miles per hour. The business was financed by a man called Sir Charles Dance, who thought that steam engines running along railways were a dead end and that the future of transport belonged to steam-powered road vehicles.

Because of the reluctance of passengers to be seated in such close proximity to the boiler of a steam engine, Goldsworthy Gurney's next model was of what he called a 'drag'; a steam locomotive which pulled a carriage behind it, rather as a coach is pulled behind a railway engine. This was a safer arrangement all round.

Just as in the twentieth century, a major factor in invention and development in the nineteenth century was the interest of the military in some development or other. During the Second World War, the building of computers was driven by such things as the need to calculate the trajectories of artillery shells and the best way to construct an atomic bomb. More recently, the Internet grew from the necessity of ensuring that a nuclear attack could not disrupt a network of linked computers. So too with the use of self-propelled steam carriages.

The very first horseless carriage was a steam-powered tractor invented in 1770 by Frenchman Nicholas Joseph Cugnot. It was intended as a tractor for hauling artillery around a battlefield. The possible military applications of Goldsworthy Gurney's steam carriage soon attracted the attention of the Army, who saw that the new means of transport might prove very handy for moving troops about the country. There was a good deal of unrest in Britain during the 1820s and finding the best way of rushing soldiers to a particular trouble-spot was one of the problems facing senior officers. Before the establishment of a proper police force, calling out the military to deal with outbreaks of rioting and disorder was a regular occurrence. Often, a garrison was 100 miles from the scene of the disturbances and marching a body of troops there could take days. The railway was used early on for this purpose. In 1839, not long after passenger services first began running in England, a train carried soldiers of the 10th Foot from Liverpool to Manchester to put down an outbreak of rioting there. Gurney's steam carriage attracted the attention of the Army almost as soon as it became public knowledge.

In the early summer of 1829, Gurney was contacted by the Quartermaster-General and invited to show what his invention was capable of. It was known by this time that his vehicles were able to move under their own steam around a city and even to make short trips to nearby villages. What was not known though was whether they would be capable of sustained, long-distance travel between two cities along ordinary country roads, essential if they were to be of any use to the army. The test of Gurney's invention was to be a journey from London to Bath and back; a round trip of roughly 250 miles by the roads existing at the time.

On 27 July 1829, the military assessment of Gurney's 'drag' was undertaken. The journey began from the Cranford Bridge Inn, which was on the Bath Road, just outside London. Driving the drag was Gurney himself, assisted by two engineers. The carriage being towed by the drag contained Goldsworthy Gurney's brother Thomas and two army officers. A horse-drawn carriage containing an extra supply of coke drove behind the drag and its carriage.

An unfortunate accident occurred when there was a collision with a mail coach, which broke one of the driving irons supplying power to the wheels. This meant that for most of the journey, it was necessary to make do with only one wheel being driven by the steam engine, which reduced their speed and power. Because of this, when they came to hills, everybody had to get out and push! Nor was that the only mishap encountered on the way to Bath. In the evening, they arrived at the Wiltshire town of Melksham, where a fair was being held. The appearance of the strange contraption was met with jeers and menacing cries.

One of the reasons that the Army were being called out to suppress demonstrations and riots was that the Industrial Revolution had caused upheaval in the lives of many ordinary workers. Steam-powered machinery had meant that much production was now being undertaken in large factories, rather than on a small scale in people's homes. This in turn forced down prices for goods and meant that a lot of previously self-employed workers were forced to work for big companies, rather than being their own masters. Anger about this situation manifested itself in hostility towards machinery, especially machinery which was powered by steam engines. Turning up at a little country town with a puffing steam engine was almost guaranteed to provoke ill-feeling. There were cries of 'Down with machinery!' and 'Knock it to pieces!' The drag and its carriage left town in a hail of stones and curses.

On the whole, the test was a success. Taken as an average, the speed on the journey to Bath and back was 15mph; twice as fast as the mail coach travelling the same route could manage, even with frequent changes of horses. The officers who travelled behind the drag must have given a favourable report to their superiors, because a month later the Prime Minister himself, the Duke of Wellington, asked for a demonstration of the drag's abilities. The duke asked that Gurney bring his machine to Hounslow Barracks in West London on 12 August. The date itself showed how much interest the duke had in the new means of travel. In the usual way of things, the 'Glorious Twelfth', the first day of the shooting season, would have seen the Duke of Wellington out on the moors with a shotgun. In 1829 though, he arrived at Hounslow

Barracks and rode around the parade ground in his coach, being pulled along by Gurney's drag. He then watched with interest as a wagon containing no fewer than twenty-seven soldiers was also towed along effortlessly.

In the event, the Army decided not to use Gurney's steam carriage, but by then his chief backer, Sir Charles Dance, had come up with a scheme which entailed using steam power to run coach services between two cities in the west of England. A regular service between Cheltenham and Gloucester began in 1831 and operated for nine months. The service ran four times a day and many men, as well as a lot of women, travelled to and from their home towns by steam carriage, just for the excitement of the trip.

Gurney's carriages had a number of features which, although novelties at the time, are now to be found in every modern road vehicle, including speed-changing gears, compensating steering geometry and differential drive hubs. Gurney really was ahead of his time and there was no reason at all why steam coaches and cars should not have replaced horses and carriages on the roads of Britain. There were, however, powerful interests at work against steam travel by road.

The companies running stagecoaches and others whose businesses relied upon horses and the stagecoach trade, such as stables, blacksmiths and inns, were vehemently opposed to the idea of steam-powered travel in any shape or form. Wherever possible, they tried to prevent or delay the opening of new railway stations, where it was felt that these might interfere with coaching. The effects of this hostility to steam power are still evident today, as one example will show.

In the early nineteenth century, the town of Kingston-upon-Thames lay on the main road between London and Portsmouth. Kingston owed its prosperity in large part to the coaching trade. Passengers stayed overnight at inns in the town, bought provisions there, changed horses and a hundred and one other things. Nobody, least of all local tradesmen, wanted to see the coaching business harmed. When the London and Southampton Railway wished to build a station in Kingston in 1838, there was widespread opposition to the move. Many people's livelihoods were threatened, and permission was refused. The railway station was built instead at Surbiton, which was at that time a tiny hamlet. It was only years later that a station was built at Kingston and because the main line now bypassed the town via Surbiton, it had to be built at the end of a spur. Today, trains for Surbiton are frequent and regular; those for Kingston, on the other hand, are anything but. This is just one example of something which was happening all over Britain at that time, as coaching interests fought a bitter battle against steam transport.

The steam carriage shuttle running between Cheltenham and Gloucester was similarly unpopular from many points of view. The coaching companies combined with the turnpike trusts, who were responsible for the upkeep of roads, to drive Sir Charles Dance's service out of business. This was done partly by sabotaging the vehicles themselves, partly by persuading people that the steam carriages were dangerous, but chiefly by arranging for the passing of an Act of Parliament which made the buses running between the two cities unprofitable. There can be no surer method of closing down any commercial enterprise than to make sure that it is no longer profitable. Once this is done, the business will usually wither away of its own accord.

Vested interests, backed by both the stagecoach companies and the turnpike trusts, succeeded in pushing legislation through Parliament which raised the toll on steam road vehicles. Stagecoaches had to pay a toll of two shillings (10p) a journey. For steam carriages, it was now set at £2, twenty times as much. Little wonder that the Cheltenham and Gloucester service folded up. It was not that the technology was faulty, simply that the tolls rendered the whole enterprise economically unviable. We shall, in a later chapter, see precisely the same thing happen to road haulage firms in the twentieth century using steam lorries, which were put out of business by the deliberate imposition of punitive taxation.

We see now how gurneys were able to flourish in the alternative world of *The Difference Engine* in a way that they could not in the real nineteenth century. In *The Difference Engine* a new political party, the Industrial Radicals, came to power in the 1830s and was dedicated to technological advancement of all kinds. The Luddites who, in our world, sabotaged steam coaches and agitated for their abolition, were, in the alternative universe, ruthlessly suppressed. Progress was encouraged and financed by the new government, rather than being treated with suspicion.

Before looking at another promising early example of steam-powered travel by road, it might be interesting to learn of a different form of transport which was being developed in both Britain and France at roughly the same time. This was an environmentally-friendly way of getting about, faster than a horse-drawn coach, and creating no pollution whatsoever, while costing literally nothing to run. One would think that these considerations alone would render this attractive in the modern world. There were, inevitably, drawbacks.

George Pocock was a schoolteacher from Bristol, who in 1800 started his own school. He was also an inventor. His first patent was for something which would have been dear to his heart as the headmaster of a boys' school. It was the Royal Patent Self-Acting Ferule, which was a machine which enabled a number of boys to be spanked simultaneously. There is no record of this ingenious aid to learning having been developed commercially and it was Pocock's next invention for which he is known today. He noticed that running a team of horses was a costly and time-consuming business and felt that the age of the horse was drawing to a close. Of course, others felt the same way, which is why steam power was coming into its own at this time.

The method of propulsion for road vehicles which George Pocock was promoting in the early nineteenth century was a renewable source of energy which has been increasingly popular in Britain over the last few years. It is difficult to drive far in England these days without catching sight of a wind turbine, but that wind could also provide the motive power for a carriage is probably a strange and counter-intuitive idea to most of us.

From childhood, George Pocock had been fascinated by kites and discovered that when he attached two large kites to a pony carriage, it was drawn easily along. After some initial experiments, he found that a wheeled vehicle weighing half a ton could be pulled along in this way by kites. Nor was the direction of travel limited or restricted by the direction of the wind. Just as a yacht can sail at right-angles to the prevailing wind, so to with Pocock's kites. He devised an ingenious arrangement of four ropes to ensure that he could control the kites without any difficulty.

In 1826 a patent was granted for the 'charvolant', for so Pocock had christened these new horseless carriages. He wrote and published a book about what he fondly hoped would one day replace horse-drawn carriages. *The Aeropleustic Art or Navigation in the Air by the use of Kites, or Buoyant Sails* was published in 1827 and in it George Pocock described the technical capabilities of his charvolants. No matter how still the air is at ground level, there is usually some movement up above and Pocock claimed that his carriages could average 20mph and even reach bursts of 25mph. This sounded then, as it does today, quite unlikely, but in public demonstrations against ordinary stagecoaches, Pocock showed that he could easily outpace them. In a race between the stagecoach from Bristol to Marlborough, for instance, the charvolant arrived in Marlborough 25 minutes before the stage. In 1828, he staged a demonstration of his carriage before King George IV. Shortly afterwards, he raced a coach running between Staines and Hounslow, once again beating the horse-drawn carriage comfortably.

On 18 July 1828, a boat was drawn across the River Mersey, against the tide and the wind, by two of Pocock's kites. For a time, it looked as though wind power might be the future of transport, but with the coming of the railways, interest in the idea of kites faded. George Pocock remained as enthusiastic as ever though. To show his faith in kite power, he strapped his young daughter in an armchair and had her drawn 300ft into the air by his kites. The child seemed to suffer no ill effects from this aerial journey and went on to become the mother of Victorian cricketing legend W.G. Grace. In France too, at about the same time, wind power was being exploited for land transport, although there the method used was slightly different.

The July 1834 edition of the *Athenaeum* magazine mentioned in passing a French experiment which had been named the *voiture à voiles l'Eolienne* and was a wind-powered carriage. The article read as follows;

> An experiment has been made at Paris, with a coach propelled by wind. It is styled voiture a voiles l'Eolienne. It started from the Ecole Militaire with a south-east wind, and reached the place Louis XV. It is stated as remarkable, that during the progress of this experiment there was a violent gust of wind, and that the carriage ascended the Pont Louis XV with a wind which was almost contrary.

The expression *l'Eolienne* means in French today 'windmill' or 'wind turbine'. It is derived from the name of the Greek god of wind. In 1834 though, the meaning was a good dealer vaguer and less precise. Images of the wind-driven carriage show that it was more similar to a large sailing ship than it was some sort of land-yacht. The sails which hung from the central mast had rigging so complicated that a crew of experienced sailors were needed to ensure that it caught the wind in the best way possible. By all accounts, it travelled much faster than any horse-drawn vehicle of the time, but there were several serious disadvantages. One of these was of course that it required so many men to keep it going. Another was that it was wholly dependent upon the prevailing weather, but the most fundamental defect was that because the sails were so large, there was no way of bringing the carriage to a halt, short of sending men into the rigging to furl the sails. Applying any kind of brake would, with the size and height of the sails, have caused the whole thing to overturn.

Road vehicles powered, at least partly, by sails feature in steampunk novels of course, most notably in *Pavane.* In addition to steam-powered lorries, the Catholic Church permits, towards the end of the twentieth century, the use of very small internal combustion engines. Because these are not adequate in themselves to power a car, they are equipped with colourful sails, to utilize the power of the wind to supplement that of the tiny petrol engines. The sails are made a feature of, being brightly coloured and gaudy, and result in such vehicles being nicknamed 'butterfly cars'.

The collapse of Goldsworthy Gurney's coach service in the west of England was not the end of steam-powered road transport. With the railways expanding exponentially across the country, it just seemed obvious to some people that if steam could carry people on rails, then it might equally well transport them from place to place along roads. The idea was very much in keeping with the spirit of the age; an exciting innovation which would revolutionize the way of life in Britain. Unfortunately, a lot of people did not want their way of life to be revolutionized; they were quite happy with things the way they were.

In 1833 a man called Walter Hancock set up the London and Paddington Steam Carriage Company. At that time, Paddington was a prosperous village just outside London. Many bankers and businessmen lived there and Hancock thought that they might wish for a new and speedier way of commuting to and from their offices each day. The year before, he had arranged for the building of a 10-seater bus, which he used to carry people from Stratford in East London to the City. He had no license to carry fare-paying passengers for this service and it was run as an experiment to gauge both the capabilities of his steam bus and also to see how the public took to the idea. All the signs were positive on both counts and so on 22 April 1833, the vehicle *Enterprise* began carrying passengers from Paddington to the City of London. Later, another bus was added to the route, the obscurely named *Autopsy*.

Over the next three and a half years, Walter Hancock's steam buses carried 12,761 passengers to the City and back, covering 4,299 route miles in the process. A third bus was added to Hancock's fleet. This was the *Automaton* and it may be seen in Illustration 7. Other companies began following Hancock's lead. In the north of England, the London, Holyhead and Liverpool Steam Coach Company began to try and attract trade from those catching or leaving ships, while the Steam Carriage Company of Scotland ran hourly services between Glasgow and Paisley.

Two things brought an end to the urban steam carriages of the 1830s. The manoeuvres of the coaching interests, which included sabotage and the flagrant rigging of tolls to penalise steam vehicles, was one factor. The other was that there was simply a huge amount of competition in the cities from the horse-drawn buses which were appearing at that time in such profusion. So many bus companies were operating on the same routes, especially the one between Paddington and the City of London, that profit margins were cut very fine indeed. Most of the companies running buses in the big cities went out of business and so did Walter Hancock and his company.

We shall return to the subject of steam transport on the roads in another chapter, but it is time now to see what was happening on the railways and, in particular, to examine the atmospheric or pneumatic railways which for a while looked as though they might be about to dominate railway transport.

We are all of us perfectly familiar with the mode of operation of steam trains. A heavy locomotive, which burns coal to operate a steam engine, pulls a line of coaches which contain either people or goods. This seems so blindingly obvious that readers might wonder why it has even been necessary to describe a steam train. What other way of managing the thing could there possibly be? Before thinking about this question, we consider a surprising fact. The fastest rail journey from the Devonshire town of Newton Abbot to the railway station in the centre of Exeter takes 32 minutes today. In 1848, this journey was covered by rail in just 20 minutes by a train which was not even drawn by a steam locomotive. Clearly, something interesting must have been going on there, 10 years after Queen Victoria came to the throne. If the journey could be undertaken then in less than two-thirds of the time that an electric train takes to cover the same distance today, then perhaps we might profit from finding out what those Victorian engineers were doing.

In the Introduction, we looked at an old print from 1829 which showed a supposed idea for travelling to India by means of a vacuum tube. This may be seen in Illustration 1. In fact, something of the sort was first mooted in the eighteenth century and the scheme appears in Victorian speculative fiction. Although hopelessly beyond the technology of the time, it was suggested that a large cast-iron pipe might be constructed and a carriage fitted within it, with an airtight seal between the sides of the carriage and the interior of the pipe. If the air in front of the carriage were to be evacuated, a partial vacuum would be created. The air pressure from behind would thus cause the carriage to move forward. It was not until the 1860s that engineers were able to put together something working on these principles, but a simpler idea was found to be a practical proposition almost as soon as railways began to be built in Britain.

In an ordinary steam train, the heaviest component is the locomotive itself. In other words, a large part of the power of the steam engine is expended on hauling itself along the track. What if the engine could remain stationary and all its energy be devoted to pulling the carriages full of people or wagons loaded with goods? Not only would this be a better and more economic use of the engine, or so the theory went, but it would also obviate many of the more disagreeable aspects of travelling by steam train. Anybody who remembers travelling by steam or who has been on a 'heritage' railway operating steam engines, will know that they are extremely noisy, smelly and dirty. Smoke and steam are an inextricable part of any journey by steam train and the racket that a large locomotive makes is appalling.

It was considerations such as those above which led to the birth of the 'atmospheric' railway. Such railways are known as 'atmospheric', because it is really the atmosphere, in the form of air pressure, which propels the train. Later developments entailed increasing the air pressure from behind. Railways of this kind are also known as 'pneumatic', because they rely essentially upon air for their movement. Here is how the first atmospheric railways were built.

In 1843 two brothers called Jacob and Joseph Samuda ran an engineering and shipbuilding firm in East London. They were also experimenting with atmospheric railways. When the greatest engineer of the Victorian Age, Isambard Kingdom Brunel, became involved with building a railway in south Devon, he decided that the Samuda brothers and their atmospheric railway would be the best and most effective way to proceed. The Samuda brothers had already done some work on the London and Croydon Railway, one of the first railways in Britain, which opened in 1839.

In 1844, the London and Croydon Railway decided that they wished to lay an experimental line which would be run by atmospheric pressure. The carriages would not run through a tube, they would be out in the open, just like ordinary railway trains. Between the tracks, a cast-iron pipe 9in in diameter was laid. This had a slot at the top, which was sealed by a leather strip. A piston was fitted into the pipe and hooked up to the front carriage of the train. When a vacuum was created in the pipe, ahead of the piston, then atmospheric pressure would drive the piston forward, towards the vacuum. The piston would thus pull the train along behind it. This sounds a strange and unwieldy way of hauling a train along and yet it worked. Stationary steam engines, housed in brick buildings at intervals along the line, created and maintained the vacuum. The trains moved silently and without any annoying smuts of soot to get on people's clothes or smoke and steam to choke them. All the smoke and steam was vented from the tall chimneys of the pumping stations, well away from the train itself.

In Ireland, the Dalkey Atmospheric Railway had opened a year before the experimental line at Croydon and also used a system produced by the Samuda brothers. This was only a short stretch of line, a little under two miles. When Brunel was deciding what sort of line to build in south Devon, he visited the

Dalkey line near Dublin to see how well it ran. In fact, because it was so short and only had one engine house, the Dalkey Atmospheric Railway ran very efficiently indeed. It was incredibly fast. Although when it began running for passengers the speed was about 40mph, the vacuum piston could actually pull a carriage along at more than twice this speed. Before services began, a young man called Frank Ebrington was inadvertently drawn along the track in a single carriage, the others not having been coupled to it. Over the two miles of the journey, the carriage travelled at an average of 84mph. There is no doubt that Frank Ebrington was the fastest man on earth for many years.

After satisfying himself that the railway at Dublin was a commercial success and that there were no snags that could be seen, Brunel threw his weight behind the use of an atmospheric railway for south Devon. That the proposed line was to be ten times as long as that at Dalkey did not seem to the great man a matter of any real importance.

The great railway engineer of the age, George Stephenson, would have nothing to do with atmospheric railways, dismissing them as 'Humbug from beginning to end'. It was perhaps because he wished to set himself as a something of a visionary, in comparison to the mighty Stephenson, that Brunel chose to make the south Devon line an atmospheric railway. The Dalkey line was very short and the one in Croydon had only ever been set up as an experiment; nobody had ever built and run a long stretch of railway on this principle.

A number of problems soon became apparent with a 20-mile stretch of atmospheric railway that were not likely to cause difficulty when only a mile or two of track was involved. One of these was that a major selling point of the railway for investors was that it would be cheaper to run than a conventional line, with carriages hauled along by a locomotive. This saving might be possible, provided that the stationary engines only operated when necessary. The original idea was that the engine houses would be situated every three miles and that an electric telegraph line would signal each of them when they needed to begin pumping to create the vacuum needed for the system to work.

Perhaps to economize, the South Devon Railway did not get around to employing enough efficient telegraph operators. This meant that instead of the pumps to create the vacuum being switched on shortly before the train was approaching, they were kept running almost continuously. This was both time-consuming for the workers and expensive in terms of the coal that was used. The seals on top of the pipes in which the vacuum was created were made of leather, which had a tendency to stiffen and crack. This meant that air could leak in and the vacuum would no longer be effective.

The difficulties of the lack of a telegraph line and the deterioration of the leather seals on the iron pipes combined with various other mishaps to cause the South Devon line to lose money for the shareholders. This was unheard-of at that time; the whole point of having shares in a railway company during the boom was that it was a sure way to riches. Little wonder then that the line closed down in 1848 and later reopened with conventional steam engines drawing the passengers along behind them. This was not the end for atmospheric railways though, far from it.

Railways which depend upon air pressure as their motive force sometimes rely just upon the creation of a vacuum at the front of the train and leaving the atmosphere behind to provide the motive power. Others blow air from behind to move it forward. Both these methods are pneumatic, in that they rely upon air to do the work. An early steampunk novel from 1972 has already been mentioned; Harry Harrison's *A Transatlantic Tunnel, Hurrah!* This tells the story of an attempt to lay an iron tube on the Atlantic seabed, create a vacuum in it and then run trains through at extremely high speed. We have seen too that this idea also featured in a short story by Jules Verne's son, which was published in 1895. Both tales foreshadowed Richard Branson's and Elon Musk's 'Hyperloops'. Surprisingly, the concept of laying a huge iron pipe under water and using air pressure to push or pull passenger trains through it comes not from either steampunk or nineteenth-century speculative fiction. The building of such a railway was begun in London in 1865 and was to be modelled upon one which was running at Crystal Palace in 1864.

In 1859, a company called the Pneumatic Despatch Company was started in London. Its aim was to dig small tunnels beneath London and use them for moving about both goods and letters, so avoiding the heavily-congested streets of the capital. To demonstrate how well such a miniature railway would work, a pipe a quarter of a mile long was laid in Battersea Park in 1861, and a stationary steam engine sucked and blew air in and out of this to propel little carriages from one end to the other. The 2ft-wide pipes ran up and down little hills which had been built from earth, the idea being, according to the *Illustrated London News*, 'various irregular curves and gradients were introduced to show that hills and valleys would not prevent the effective working of the system'. This was, in effect, what we would today call a 'publicity stunt', designed to attract public attention to the plans of the Pneumatic Despatch Company. The company not only had plans for moving freight by means of a pneumatic railway; they were also thinking about the possibilities of passenger transport.

The Post Office was interested in the idea of an underground railway, not to carry passengers but rather to send sacks of mail from one sorting office to another. The Pneumatic Despatch Company was commissioned to build a trial section of tunnel under ground, to see how well this method would work in practice. To give some idea of how a journey by this system might feel, volunteers offered to take a ride. According to the 24 August 1861 edition of the *Illustrated London News*;

> Two gentlemen occupied the carriages during the first trip. They lay on their backs, on mattresses, with horsecloths for coverings, and appeared to be perfectly satisfied with their journey.

Apart from finding the 2ft-wide carriage a little cramped, the passengers remarked upon how pleasant it was to travel without noise, steam and soot. According to the *London Journal*, a woman too made the journey in a similar way;

> Not only have letters and parcels been transmitted through the tube but we hear also that a lady, whose courage or rashness – we know not which to call it – astonished all spectators, was actually shot the whole length of the tube, crinoline and all, without injury to person or petticoat.

After this demonstration, the Post Office engaged the Pneumatic Despatch Company to dig a tunnel from Euston Station to the Post Office's North Western District Office, 600 yards away. This opened in February 1863. The idea of a passenger railway through tubes had captured the imagination of those running the Pneumatic Despatch Company though and three years after the demonstration of the small-gauge pipe in Battersea Park, one of the men who had set up the tube for the post office, followed it up by building a short section of a full-size passenger-carrying train, which was constructed in the park surrounding the Crystal Palace.

In the summer of 1864, a tunnel consisting of a pipe with an oval cross-section, about 10ft high and 9ft wide, covered with earth, was set up, running from the Sydenham Entrance to the park all the way to the Penge Gate; a distance of a little over a quarter of a mile. It may be seen in Illustration 2. The tunnel entrance was made of brickwork and a powerful steam engine drove a fan, which both sucked air from in front of the train and blew it from behind. This method had enabled the carriages in Battersea Park to reach a speed of about 40mph, but nobody wished to see a carriage containing thirty-five people travelling that fast. The scope for a disaster over such a short distance was obvious if the single carriage had shot out of the end of the pipe at such a high speed.

The pneumatic railway at Crystal Palace was little more than a fairground ride; it was simply a new experience for those with 6d to spare. The reaction of the public was very favourable indeed. The ride in the carriage, which was described in one newspaper as resembling an 'elongated omnibus', was smooth, silent and clean. Many of those who rode through the tunnel compared it favourably with ordinary railway travel. Despite this, the line ran only for two months, carrying its last passengers on 31 October 1864. Just as the experimental line in Battersea Park had led to commercial work for the Post Office, so too did the set-up in Crystal Palace Park result in a company beginning work on a full-sized project; in this case the construction of a new underground railway line which was to pass under the Thames.

Before looking at the first attempt to build a full-scale, commercially-viable atmospheric railway, readers might be interested to know that rumours persisted for over a century after it ended running that the remains of the Crystal Palace atmospheric railway, not only the tunnels and tracks but also the carriage itself, had simply remained where they were after the line had stopped running and been covered over with earth, rather than the company going to all the trouble of dismantling and removing the infrastructure. It is quite possible that this story arose because that is precisely what happened to another short stretch of atmospheric railway which was running at about the same time in New York. This was indeed simply abandoned and rediscovered many years later.

In August 1975 an excavation by the London Underground Railway Society was carried out at Crystal Palace and unearthed a flooded brick-lined tunnel. In 1989 a German archaeologist found the remains of some of the track. Perhaps the most bizarre story is that of a woman who was walking in the park in 1978. She claimed to have fallen down a shaft and found herself in a dark, subterranean chamber. When she lit a match, she said that she had found to her horror that she was next to a mouldering railway carriage, which contained the skeletal remains of passengers dressed in Victorian clothes. She scrambled out, but nobody was ever able to find again the hole down which she had supposedly tumbled.

We return now to the building of a commercial atmospheric railway; one which would actually be a profit-making enterprise, rather than merely a pleasure ride. The mid-1860s were the perfect time to be starting an underground line in the capital. London's, and the world's, first underground railway had opened in 1863 and proved immensely popular with Londoners. On the day it opened, 10 January 1863, the Metropolitan Railway carried 38,000 passengers. In the first year of operation a staggering nine million people travelled on the underground railway; no mean feat in a city which had at that time a population of fewer than two million. After seeing how many passengers were using the Metropolitan Railway and bearing in mind the great interest shown in the short section of line at Crystal Palace, it is little wonder that a company was established which would construct the second underground line in London. The prospectus of the Waterloo and Whitehall Pneumatic Railway Company was issued on 15 May 1865.

The advantages to customers of travelling by a pneumatic underground railway were considerable. The ordinary steam engines running in the tunnels of the Metropolitan Railway created a truly terrible atmosphere. The smell was appalling and although steam was vented when the engines were passing a point which was open to the sky, there was still a good deal of steam and smoke for passengers to contend with; to say nothing of the noise. All this was bad enough out in the open, but in underground tunnels it was most unpleasant. A pneumatic underground railway would do away with all that. As the prospectus had it;

> From the absence of smoke, steam, and other objectionable accompaniments of the locomotive, and the complete ventilation of the tunnel by the continuous draught of air through it, the working is attended with perfect comfort to the passengers.

The new line would run from Waterloo railway station, under the Thames and then emerge near Scotland Yard, just behind Whitehall. Commuters arriving at Waterloo from the suburbs would not have any distance at all to walk to offices in the area of Whitehall, nor would they have to pay the halfpenny toll to cross Hungerford Bridge to get to the other side of the river where Charing Cross Station now stands. The capital to be raised was £135,000, divided into £10 shares. It was a certain winner. A bill was passed through Parliament, allowing the building of the new railway and work began on 25 October 1865. It was estimated that the whole of the work would be completed within a year.

Some of the technical details of the pipeline being laid across the bed of the Thames are very similar to the descriptions of the transatlantic tunnels described by both Harry Harrison and Michel Verne in 1972 and 1895 respectively. Harry Harrison has the protagonist of his novel talk in these terms of the theoretical speeds which his train in a vacuum tunnel might attain;

> With the air removed we can now consider higher speeds than were ever possible before. Why, there is no reason why our trains cannot go eight, nine hundred – even a thousand miles an hour.

These are the kind of speeds which Richard Branson and Elon Musk have been considering for their own projected vacuum tubes and are incidentally, similar to the speed which Michel Verne talked of in his science fiction story. Here he is, writing about more of the specifications for the transatlantic tunnel through which he was to travel from Liverpool to Boston, in *An Express of the Future*;

> Complaisantly the journalist entered into the details of the enterprise. He stated that more than 3,000 miles of iron tubes, weighing over 13,000,000 tons, were required, with the number of ships necessary, for the transport of this material – 200 ships of 2,000 tons, each making thirty-three voyages.

Similarly massive weights of iron tubing would be needed to cross the Thames. There would be four iron tubes, each of them 250ft long and weighing 1,000 tons apiece. This engineering project though was not science fiction or steampunk; construction actually started in late 1865 and continued for some time.

Work on the parts of the railway which would be on dry land, at both the Whitehall and Waterloo ends of the line, went ahead and were, by June 1866, progressing well. The massive iron tubes needed for the river crossing were being made at the works of the Samuda brothers, the same company involved with the South Devon Railway. There was some debate as to whether they should be welded together on site and then floated down the Thames to Whitehall or brought individually and then joined up on the river bed. Everything looked as though it were going smoothly, when there occurred one of those unexpected events which casts a very long shadow. The most respected bank in Victorian London, Overend, Gurney & Co, known as 'the banker's bank', collapsed. The shockwaves from the failure of this one bank rippled outwards and saw hundreds of banks and other businesses also fail.

Following the financial crisis caused by the Overend, Gurney & Co affair, a number of railway companies also collapsed. These were businesses which looked perfectly sound but were actually overstretched financially. The Waterloo and Whitehall Pneumatic Railway Company found that there were one or two small expenses which they had not bargained for, but they could easily be solved with the injection of an additional £115,000 of capital. Unfortunately, no banks felt like lending money on the enterprise. An American company showed interest, an attempt was made to borrow the money from another railway, but it was all to no avail. The grand scheme simply ran out of money and there was no more to be had, at least not for a new and untried method of propelling trains. The company collapsed and this was almost the last that was heard of pneumatic railways in Britain.

There were one or two attempts to build pneumatic railways after the failure of the Waterloo to Whitehall line. A few years later, plans were laid for a short line running from South Kensington Station to the Royal Albert Hall, but nothing came of it. Instead, in 1885, a pedestrian foot tunnel was dug along roughly the same route that the pneumatic railway would have run. This exists to this day, although there were calls in 1887 for it to become part of an extension to the Piccadilly underground line. There was to be a coda to the saga of pneumatic railways, but it was played out not in Britain, but America.

Alfred Ely Beach was an inventor and he had been impressed by what he had heard of both the Crystal Palace experimental pneumatic railway and also the failed project to run a pneumatic railway under the River Thames. In 1867, shortly after the Waterloo to Whitehall railway was acknowledged to be dead, Beach demonstrated a model pneumatic railway at the American Institute Exhibition held in New York that year. That same year, he set up the Beach Pneumatic Transit Company and began excavating a tunnel beneath Broadway, which would be the first section of a pneumatic subway system. There was only one small difficulty about this project; he had no planning permission to build an underground line and so the whole thing would have to be undertaken in secret.

It sounds like a fantasy novel in itself; the story of a man who tried to excavate a tunnel under one of the world's busiest and most populous cities and build an underground railway, without anybody noticing what he was up to. However, New York's first subway really was established in secret and it was not until the day it opened that most of those living in the city had the faintest idea what had been going on under their feet. This is a tale which really deserves a book of its own, rather than just a brief mention like this.

Alfred Ely Beach had obtained planning permission for running two underground pneumatic postal tubes beneath New York, similar to those upon which the postal system in London once depended. Once he had permission for an experimental stretch of two small tubes, he began digging in earnest an 8ft-wide tunnel, which ran for 312ft beneath Broadway. He had to work very swiftly, before the authorities found what he was up to. The whole project was completed in less than two months.

A single carriage ran between two luxurious stations, which even featured a piano in the waiting room. Air pressure drove the carriage from one station to another and the whole thing was really a novelty or trial run for a more extensive system rather than a practical way of getting around the city. In the first year of operation, 400,000 passengers paid 25 cents each to be transported silently from one dead-end station to another. It was a pleasant experience, with none of the steam, smoke, dirt and noise which were inextricably associated with railway travel at that time.

Having presented New York's City Hall with a *fait accompli*, Alfred Beach sought investors and applied for planning permission to extend his underground atmospheric railway to other parts of the city. Three years later, official permission was finally granted, by which time it was too late. Just like the Waterloo to Whitehall underground railway in London, Beach's transit system fell victim not to technological difficulties, but economic ones. A stock market crash caused the collapse of his company and no more tunnels were excavated. Just as is the case with the Crystal Palace atmospheric railway tunnels, there are rumours that a section of Beach's tunnel still exists and may be accessed from a manhole cover near Broadway. In 1912, the carriage was certainly discovered during excavations for the present-day New York underground system.

Steam-powered road vehicles and railway trains which run through tunnels from which the air has been evacuated are two of the strange ideas which not infrequently crop up in steampunk. Another is some means of speedy communication which does not require the use of electricity. In the next chapter, we shall be looking at a device which has appeared in many books of the genre at which we are looking; most notably in the *Discworld* books of the late Sir Terry Pratchett.

Chapter 3

The Mechanical Internet

I t is sometimes claimed that before the advent of the electric telegraph, at about the same time that Queen Victoria came to the throne, information could be transmitted no faster than a galloping horse or a sailing ship. Consider though the following incident, which occurred 40 years before the first practical electric telegraph was introduced. In the late eighteenth century, conditions on board ships of Britain's Royal Navy were appalling. The food was bad, living quarters dreadful and the discipline harsh and unremitting. Mutiny was a constant fear among the officers of the navy; the prospect that their men would rise up against them and take over the ship, as had happened during the mutiny on the *Bounty* in 1789. In the spring of 1797 HMS *Royal Sovereign* was moored in the southern English port of Portsmouth, near that part of the channel of sea between the Hampshire coast and the Isle of Wight which is known as Spithead.

A sailor on HMS *Royal Sovereign* told one of his officers privately that a mutiny was indeed about to break out on the ship. The warning was taken seriously, for the men of this and other ships had lately drawn up petitions, which they had sent to the King, complaining about conditions on board the ships of the Royal Navy.

Unrest and rebellion can be contagious and so once it was known that one ship might be on the point of mutiny, it was vital to seek the advice of the Admiralty in London, so that orders might be received regarding the best way to deal with such a dangerous situation. On 1 April 1797, a signal was despatched to London as a matter of urgency. It was brief and to the point, reading, 'Mutiny brewing in Spithead'. The astonishing thing is the rapidity with which this brief message travelled the 70 miles to London. It took just three minutes from sending it until it was being read in London. The signal had travelled between the two cities at 1,400mph or twice the speed of sound.

In the world of steampunk, mechanical devices regularly replace those which, in our own world, run on electricity. The computers in *The Difference Engine*, for instance, are composed of gleaming cogs which are spun by steam engines rather than the electrically-driven microprocessors which we usually associate with computing. Long-distance communications in steampunk fantasies tend to rely too upon systems which use steam, clockwork or manpower, rather than a supply of electricity. The *Discworld* novels of Terry Pratchett contain a perfect example of this sort of thing.

Although the *Discworld* novels began as pure fantasy, they gradually developed over the years until some aspects of them were indistinguishable from steampunk. Indeed, the fortieth book in the series, published in 2013, is called *Raising Steam* and actually features a steam engine. In earlier *Discworld* books, a chain of mechanical signalling stations make an appearance. The 'clacks' are rickety wooden towers which have at the top a display of shutters. These can be either open or closed, and the various combinations are used to encode messages. The signal stations are on flat ground, positioned about eight miles apart from each other. Although when they are first mentioned, the clacks are really a type of telegraph system, they later evolve into something like a rudimentary Internet. As the system grows, it becomes automated, with information being stored on punched tapes. A clockwork mechanism is also used, another popular means of powering devices in steampunk stories. So sophisticated did the clacks become, that they were ultimately able to transmit not only text, but also images.

In *Going Postal*, we are told that:

> On the Tump, the old castle mound across the river, the big tower, one end of the Grand Trunk that wound more than two thousand miles across the continent to Genua, glittered with semaphore.

This is indeed, or so it would at first seem, an extravagant fantasy; a network of semaphore towers which stretches across an entire continent.

The 'clacks' were first mentioned in Pratchett's *The Fifth Elephant*, which was published 10 years after *The Difference Engine*, and it has been speculated that his use of the word 'clack' was inspired by the earlier steampunk work; those operating the mechanical computers in *The Difference Engine* being known as 'clackers'. It is equally possible that both words are derived from a common source. The expression 'hacker', in relation to computers, was already current when *The Difference Engine* first appeared in 1990.

Another imaginary world in which mechanical or optical telegraphs feature prominently is *Pavane*, by Keith Roberts. In *Pavane*, late twentieth and early twenty-first century Europe is wholly reliant upon such a method for rapid communication between cities. The Catholic Church rules supreme and has forbidden the use of electricity, branding it heretical. Instead of telephone and radio, a network of signal towers stretches across the world.

These are fitted with large semaphore arms. Each tower is within view of two others and when one sends a message by semaphore signals, a man with a pair of field glasses reads it out loud to a companion who is operating the arms of their own tower. In this way information about all sorts of things, ranging from the price of grain in London to an urgent appeal for military assistance, can be speedily transmitted across Europe and beyond.

The semaphore towers of the Guild of Signallers in *Pavane* are a thread running through the whole book. One chapter is devoted solely to this strange and seemingly primitive communications network, but it is mentioned in passing throughout the rest of the narrative. We are given to understand that even the Church relies heavily upon the signallers to pass messages around the world. There is simply no other means of rapid communication.

All that has been described so far sounds like typical, fictional alternative worlds. The idea of sending complex information hundreds of miles by semaphore or an arrangement of shutters opening and closing really does not sound like a practical proposition. No wonder we know it only through the works of authors such as Terry Pratchett and Keith Roberts. It may therefore come as a surprise to learn that this mechanical Internet did once exist in our own world and that the accounts found in *Pavane* and *The Fifth Elephant* are no more than the sober, historical truth. Two hundred years ago, a chain of semaphore towers did indeed straddle a continent; the continent of Europe. It was possible in the early nineteenth century to send a message swiftly from one end of Europe to the other by this means. The system ran from Amsterdam in the north to Venice in the south. More surprising still, the code used in some of these towers provided the basis for the transfer of written text across the Internet today.

For almost the whole of recorded history, it has been impossible to transmit complex information other than by means of a physical message in the form of either a person carrying information in his head or by a written letter. Of course, beacons can be lit and a chain of such signal-fires can rapidly convey news of some event. This is known from classical times, when the Persians were fighting against Athens and news of the outcome of a battle was sent hundreds of miles in a relatively brief period of time. More recently, beacons on English hilltops carried news from Devon to London of the arrival of the Spanish Armada off the coast in 1588.

The disadvantage of signal fires is that they effectively carry only one piece of information; which has to be arranged beforehand. They are binary, having only two positions; on or off. Once a beacon in the West Country was ignited in 1588, somebody 10 miles or so away would see it and light another fire. In this way, the signal could be passed swiftly across the land. It contained no useful information though, other than that which had previously been agreed upon. To tell the people in London how many ships had been spotted and of what types would have needed a letter to be carried on horseback for hundreds of miles. At least a couple of days were required to carry a message in that way. For thousands of years, that is how it was. Sending word from one place to another was a maddeningly slow process. As late as 1860, it took 24 days to get a message across America, from New York to California. This was because most of the distance was served only by mail coaches.

Of course, sound travels considerably faster than a galloping horse and a message can be carried on waves of sound which take just under five seconds to travel a mile. Here too though, is the same problem that we saw with signal beacons. A loud sound can carry for miles but cannot relay anything other than a pre-arranged message. After the fall of France in 1940, for example, the ringing of church bells in Britain was banned by the government. It was made known that if church bells began to sound, this would be the signal that the invasion of Britain by Germany had begun. Again, this is a single binary code. Either the bells ring or they do not.

In the late eighteenth century, a Frenchman called Claude Chappe came up with an ingenious method which would allow him to send detailed information by sound waves which he would generate simply by banging an old copper cooking pot. He arranged with his brother Rene, that they would both be looking at two specially-adapted clocks which Claude had devised. There were no minute or hour hands on these clocks, just a second hand, which had been tinkered with until it swept around the face at twice the usual speed.

To compensate for the time taken for sound to travel, the following procedure was adopted. With his brother several hundred yards away, Claude Chappe waited until the second hand on his clock was on twelve and then banged his cooking pot very hard. As soon as his brother heard the sound, he started his own clock, with the hand on twelve. In this way, the two clocks were synchronized. It was now possible to send signals across a substantial distance. A code was devised, whereby numbers were used to signify the letters of the alphabet. As soon as Rene heard the clang of the cooking pot being whacked, he noted down the number which his second hand was passing. After a sequence of numbers had been transmitted in this way, he consulted a book in which he had the code written down and then knew what his brother was saying.

Claude Chappe's system was an ingenious one, but it had obvious limitations; the chief of which was the distance that a loud noise could travel. One way of tackling this would be to make a louder sound, but it occurred to Chappe that a visual signal might be more effective. At first, he used a post with a panel which could be swiftly turned one way or the other. One side was painted white and the other black. Using the same clocks and code, Chappe found that he could now send a message much further, especially when the person receiving the signal was equipped with a telescope. On 2 March 1791, the Chappes used the new method to send a message dictated by an objective observer over a distance of 10 miles.

For most of us, the word 'telegraph' is inextricably linked with the electric wires used in the nineteenth century to carry messages in Morse code. It is interesting to consider the origin of the word and find that telegraphs were being talked about and used decades before Samuel Morse came up with his code. Indeed, the word was coined in 1791, the very year that Morse was born. When he successfully demonstrated his system over a long distance, Claude Chappe wished to find a name for his invention; something catchy and memorable. At first, he favoured 'tachygraph', Greek for 'fast writing'. A friend persuaded him that 'telegraph', which means 'far writing', would be a better name and so the telegraph was born.

Had Chappe's telegraph been invented just a few years earlier, it is altogether possible that nothing would have come of it. France was notorious for being the most hidebound and reactionary country in Europe, with no need for new ways of doing anything. However, the country was in the throes of revolution when Claude Chappe first demonstrated his telegraph and the time was ripe for radical ideas and new schemes. In 1793, the same year that Louis XVI went to the guillotine, the National Convention, the new body ruling France, set up a commission to investigate Chappe's work and see if it could be of any use to the state. The commission suggested that money should be allocated to test the idea and see if the telegraph really could speed up communications across the nation, in the way that Claude Chappe claimed that it could.

The swivelling black and white panels and synchronized clocks had proved to be a laborious and cumbersome way of spelling out words, so Chappe had teamed up with a clockmaker to produce a semaphore system which could be operated by one man. A tall mast bore a horizontal beam at the end of which were two small arms, which could move independently. The whole array could be operated by levers from ground level. The main crossbar could be tilted and the smaller bars could each be moved to one of seven positions, meaning that the whole array could adopt a total of ninety-four separate and distinct arrangements – more than enough to signal numbers, letters of the alphabet, punctuation and a number of common syllables.

Three telegraph stations were constructed between Belleville, on the outskirts of Paris, and Saint-Martin-du-Tertre, which was about 20 miles away. On 12 July 1793, a message was sent from one end of the line to the other; a process which took a little over 10 minutes. It was a fantastic achievement and the head of the commission which had been set up by the National Convention to look into telegraphy declared that this was a triumph for the French people and that henceforth France would be the nation to instruct Europe.

The real reason for the National Convention's interest in building telegraph stations was purely pragmatic. With the country in turmoil, it was vital that the government in Paris could communicate swiftly with any part of France. Mutinies and insurrections were brewing in various districts and towns and strong leadership from the central government was essential if control was to be exerted over the entire nation. We saw earlier that in 1860, shortly before the start of the American Civil War, it took 24 days to get a message from one coast of America to the other. This meant that if the government in Washington wished to know what was going on in San Francisco, it would take 48 days simply to send a question and receive a reply. Under such circumstances, it was inevitable that some parts of the United States began to think of themselves as not being closely bound to the federal government, thousands of miles away. This state of affairs contributed to the tensions which led to the American Civil War. The revolutionary government in Paris had the same problem. They did not want it to take days to be able to find out what was going on in another part of the country; this sort of delay tended to make outlying towns feel that they were somehow independent from the capital and its authority.

It took nine months to construct the first section of the French State Telegraph network, which ran 150 miles from Paris to Lille in the north, near the present-day Belgian border. It was not long before the British became interested in the idea of telegraph lines. In their case, it was less about holding the country together under the reins of a central government, than it was establishing rapid communication between London and the naval bases on the south coast.

The first British telegraph network is worth examining in detail, because it is more than an historical curiosity. The method used by this telegraph linking London and Deal gave rise to the system which we use today for sending text around via the Internet. It was a digital information network, almost identical to the ASCII code that computers now use to store and transmit written words in the twenty-first century.

It was, surprisingly, a clergyman who invented the telegraph that the British Admiralty began building in 1795. George Murray was, in addition to being a man of the cloth, an enthusiastic amateur scientist and also a lord. He was the second son of the Duke of Atholl. Murray's idea did not entail the use of semaphore arms, but rather an array of six shutters, each about 5ft high and with rods in the middle so that they could be rotated horizontally. Such an arrangement may be seen in Illustration 8. In this way, each of the six shutters could be in one of two positions: either open or closed. This is of course a binary system, where things can be either one or zero, open or closed, on or off. It is the way that modern, digital devices work.

The array of shutters was capable of sixty-four combinations; more than enough to encode letters and numbers. So the letter 'A' was signified by;

$$- \quad 0$$
$$0 \quad 0$$
$$0 \quad 0$$

'B' was:

$$0 \quad -$$
$$0 \quad 0$$
$$0 \quad 0$$

and so on. It will be readily apparent that this can be expressed simply as a string of ones and zeros. Letter 'A' would be I00000, 'B' would be 0I0000. This is of course precisely how we encode text in the modern world. We use a universal protocol called ASCII, the acronym being pronounced 'ask-ee' and standing for American Standard Code for Information Interchange. Just as with the Reverend Murray's shutter telegraph, letters and numbers are reduced to strings of binary digits; 'A' is 1000001, 'B' is 1000010. In this way, all written material can be turned into sequences of digital information and sent through wires, along fibre-optic cables or through the air by radio waves and then reconstructed at the other end. Computers can also manipulate information in this form. It was an Anglican clergyman in the late eighteenth century who first used a binary code like this for practical, long-distance communications.

We return to Terry Pratchett's 'clacks' for a moment. It will be recalled that the continent-wide chain of optical telegraph towers which ran 2,000 miles from the city of Ankh-Morpork to distant Genua was supposedly capable of carrying not only text, but also pictures. It is not hard to see how this might be done and is indeed, in theory at least, quite possible. Just as individual letters are reduced to binary digits when being sent across the Internet, so too are photographs and other images. In modern computers, these binary digits are usually manipulated in groups of eight; a byte. One byte is typically used to convey a single character. George Murray's system used six binary digits for each character, but other than that, the arrangement is precisely the same. It would be a lengthy and time-consuming process to undertake, but a picture could be broken down into units of six digits and encoded, then sent along a mechanical telegraph line. The fantasy of *Discworld*, at least in this case, is not all that far-fetched.

The first telegraph line established by the Admiralty began to be set up in 1795 and started operation the following year. It ran for 70 miles from London to Deal on the Kent coast. There were fifteen stations on the route. Because there had to be a clear line of sight between one station and another, the telegraphs were, where possible, situated on high ground or hills. There are a number of places in south-east England known as Telegraph Hill for this reason; high ground, where once an optical telegraph stood. In London for example, Telegraph Hill Park, in the New Cross Gate district, is named after the shutter telegraph erected on the ridge which rises 150ft above central London. Even with all the development of the area during the Victorian era, there is still a clear view across the whole of London from this point.

To this very day, a curious surviving relic of the Admiralty telegraph set up in 1795 may be seen in London. Illustration 17 shows No. 36 West Square, in the south London district of Lambeth. This was the first telegraph station from the Admiralty in Whitehall and, as may be seen, it differs architecturally from neighbouring properties. It has a brick-built top storey, rather than the tiled affairs which are seen on either side, and also an imposing bay window. This house was owned by the Inspector of Telegraphs and when first acquired, it was indistinguishable from all the other houses in West Square, which were built in 1794.

The Admiralty bought the lease on No. 36 and erected one of Murray's wooden shutter frames on the roof, from where there was a clear and unobstructed line of sight to the Admiralty in Whitehall. For the next 16 years, everything went smoothly. Then, in 1812, work began on the new

Bethlem Royal Hospital, which was to be built in nearby St George's Fields. The architectural plans for this lunatic asylum, which was to become known colloquially as 'Bedlam', included a dome above the entrance which would, by ill fortune, obscure the line of sight between the Admiralty and the telegraph station in West Square. At first, the Admiralty suggested an ingenious plan. If they could be allowed to rent the roof of the new asylum, then the shutter could be erected there. Consideration was given to this idea, but ultimately rejected on the grounds that the constant clattering from the operation of the shutter could aggravate the condition of some of the more sensitive patients.

There was nothing for it but to adapt the property at No. 36 West Square. A brick base was constructed on the roof, upon which a wooden platform was placed, which raised the shutter so that it would be visible from the Admiralty even when the new hospital was built. At the same time, an imposing bay window was built to show that this was no common house, but the headquarters of the telegraph chain. The address is, over 200 years later, still known as Telegraph House. The Bethlem Royal Hospital too, remains a few hundred yards away, although it now houses the Imperial War Museum.

Fifteen telegraph stations linked the Admiralty in London with the port of Deal, which lies about 70 miles, as the crow flies, from the centre of the capital. Before the building of the telegraph line in 1795 and 1796, it would have taken a day's hard riding to carry a message between London and Deal. Even then, it would have been a remarkable horse which could cover the distance in the space of 24 hours, because by road the distance is closer to 90 miles than 70. It is now that we encounter for the first time the truly astonishing speed at which optical telegraphs were able to transmit complicated information from one place to another; which might be many miles away. It took a mere 60 seconds to send a simple signal from London to Deal.

We pause for a moment to consider the implications of the fact that a signal could travel the 70 miles which separates London from Deal in just one minute. A simple calculation reveals something which leaves one gaping, open-mouthed, in disbelief. If a signal travels 70 miles in one minute, then it must be travelling at a speed of 4,200mph, more than six times the speed of sound! For the eighteenth century, this was, to say the least of it, quite a feat. Even more amazing is that this was by no means the top speed for the spreading of information by optical telegraphs. A few years later, another telegraph line in the north of England was sending information five times as fast as this.

The telegraph line from London to Deal needed fifteen stations, each staffed by four men; a lieutenant, a midshipman and two assistants. Two were employed in watching the next stations in either direction, a third wrote down messages as they were received and the fourth man operated the shutters, so that the signals could be passed to the next in line. This meant that sixty men were occupied full-time, because the spotters had to be on duty throughout the hours of daylight, as nobody knew when a message might come through. After the success of the Deal line, others were constructed. Dover, Sheerness, Great Yarmouth, Portsmouth and Plymouth all had their own lines connecting them to London. It was possible to send a message from the Suffolk town of Great Yarmouth, via London, all the way to Devon. By 1815, there were sixty-five of these telegraph stations dotted about on the hills of southern England.

Maintaining the shutter telegraph network which spread its tendrils across England in the late eighteenth and early nineteenth century was only made possible by the peculiar conditions which existed at the time. Between 1792 and 1815, Britain and France were at war almost continuously. The French Revolutionary Wars of 1792 to 1802 segued seamlessly into the Napoleonic Wars, which lasted from 1803 until 1815. The Napoleonic Wars have been called the first real world war, as the fighting took place from Scandinavia to South Africa, from the Middle East to the Caribbean and from Russia to the Indian Ocean. During that time it has been estimated that one-sixth of British men were enrolled in the armed forces and the British economy was effectively on a war footing for decades.

For over 20 years, Britain fought France and at one stage, there was even the threat of invasion from the continent. That being so, it was vital to have rapid and reliable communications between the ports from which the Royal Navy was operating and the Admiralty in London. The telegraph network which began to be set up from 1795 onwards was no luxury, but an essential part of the country's defences. The peculiar circumstances of there being so many men enlisted in the army and navy also made it possible to operate the shutter telegraph cheaply, since wages were not a consideration. The 65 telegraph stations needed over 250 men to run them which, in the usual way of things, would have been an expensive business. When, however, you already have hundreds of thousands of men under arms, it makes life far easier; you have virtually unlimited manpower available for building and operating something of this sort. Telegraphs of this sort, just like the one first set up in France, were never likely to be economically viable in peacetime.

There were though those who could see an exciting future for the telegraph; once it became a commercial, rather than a purely military, enterprise. The 1797 edition of the *Encyclopaedia Britannica* suggested that one day the optical telegraph might be run like the ordinary postal service and that people would pay to send their messages along it. It would then become a self-supporting service. The encyclopaedia also foresaw diplomatic and political advantages, if the telegraph network grew and spread further across the world, saying that;

> The capitals of distant nations might be united by a chain of posts, and the settling of those disputes which at present take up months or years might then be accomplished in as many hours.

When Napoleon Bonaparte seized control of France in a *coup d'état* in 1799, he inherited, among other apparatus of the state, the State Telegraph Service. He quickly realized the importance of this long-distance communication network and had it extended, until it stretched beyond the borders of France. It was Napoleon's hope that the telegraph could even link Britain with continental Europe. At first, he authorized the building of a new line south, stretching as far as the Italian city of Milan. Then, when he was planning the invasion of England, he gave thought to setting up semaphore towers which might one day send messages across the Channel. One was built at Boulogne, but when the invasion was cancelled, the scheme collapsed. It is intriguing to think that as early as 1805, Britain was nearly linked up with a continent-wide communications network.

We have seen that the chains of signal towers in books such as *Pavane* and *Going Postal* are not fantasies at all, but really a slightly exaggerated account of historical fact. Long before the electric telegraph, the mechanical telegraph ran across England and much of Europe. The French and British lines were not the only ones operating. In retrospect, the optical telegraph might seem to be a blind alley, but new lines were being built well into the 1850s, long after the electric telegraph had proved its efficiency. Optical telegraphs were especially popular in Scandinavia; Denmark, Norway, Sweden and Finland all constructed telegraph lines of this sort. Prussia and Russia too had telegraphs. The Russian line ran between Moscow and Warsaw and was the longest ever built, at over 700 miles. Some of these networks used semaphores, others shutters. None of the systems were compatible with each other, which made it impossible for the different countries' telegraph lines to connect with each other. Different numbers of shutters were used and various kinds of semaphore arms.

In both Canada and the United States, mechanical telegraphs were being built in the 1820s and 1830s. One line ran from Staten Island to

Manhattan and another linked Martha's Vineyard with Boston. Even Australia embraced the technology, the first line being opened in Sydney in 1827. It seemed that the mechanical telegraph was here to stay and every year saw innovations in existing systems, as well as completely new types of shutter or semaphore post.

The optical telegraph never developed into a connected, worldwide network, as the electrical telegraph did. There were several reasons for this. First, this form of telegraph had its origins in military and political necessity. It did not begin as a commercial enterprise and because it was so costly in manpower, those running the systems saw no reason to share them with others. Claude Chappe had a vision of his invention being used to carry news bulletins from one end of France to the other and also to advise on commodity prices, to bring about a unified European economy. There was no appetite in the government though for such idea, although the winning numbers of the state-run lottery were eventually allowed to be sent across the country in this way. This greatly reduced fraud.

There was no possibility in Britain either of the telegraph lines being used for anything other than military traffic. The Admiralty had built them, their men were running the things; what reason would they have for sharing them with anybody else? Other countries were a little more relaxed about this and things like weather forecasts, news and commercial information began to travel at high speed around Europe. News that had once taken a week to reach a capital city could now be there in less than an hour!

After the Battle of Waterloo, the final defeat of Napoleon and end of the wars fought between the British and French, the Admiralty in London decided that their spider's web of telegraph lines were surplus to requirements; a luxury which was no longer needed. The whole system was therefore dismantled. Only one line was wanted in peacetime and this was a link between London and the main British naval base at the Hampshire city of Portsmouth. For this, a new method was used. The shutters had worked well enough under ideal conditions and perfect visibility, but semaphores were more easily seen in gloomy or hazy weather. This was of course one of the great problems generally with optical telegraphs; the fact that in rain or fog, it could be hard to see with perfect clarity for eight or ten miles. For the British this was a particular difficulty, because of course their lines all ran to coastal ports and fog is more common near the sea than it is inland. For as many as 165 days a year in Britain, the weather is such that it is not possible to see clearly for 10 miles; even with a telescope or pair of field glasses. The London to Portsmouth line was converted to semaphore.

Throughout the 1820s there was something of a boom in the building of telegraphs and in Britain a line was built which demonstrated just how fast an optical telegraph could be. In the early 1820s, Liverpool was a major port in England. Ships simply arrived unexpectedly and had to be directed to harbour as they came. This was a chaotic and unsatisfactory situation, which led to traffic jams of vessels lining up to be allocated permission to enter the port. If only the port authorities had some way of knowing when ships would be arriving. Since most of them sailed north through the Irish Sea to reach Liverpool, skirting the coast of north Wales in the process, it was thought that a telegraph line running from Holyhead, on the Isle of Anglesey, to Liverpool itself, might solve the problem.

In 1825 the Liverpool Dock Trustees managed to get an Act of Parliament passed which authorized them to;

> establish a speedy Mode of Communication to the Ship-owners and Merchants at Liverpool of the arrival of Ships and Vessels off the Port of Liverpool or the Coast of Wales, by building, erecting and maintaining Signal Houses, Telegraphs or such other Modes of Communication as to them shall seem expedient, between Liverpool and Hoylake, or between Liverpool and the Isle of Anglesey.

The following year, twelve signal stations were built, with living accommodation for those who would be operating them. Ships' masts 50ft high were erected and three pairs of semaphore arms attached to them. The line of telegraph stations ran for 72 miles along the Welsh coast, via Rhyl, Colwyn Bay and Llandudno. An article in the *Shrewsbury Chronicle* on 21 October 1836 detailed precisely how quickly a message could be sent along the 72 miles of the line.

According to the logbook kept in the Liverpool station, it was the custom to send a message to Holyhead at exactly 1.00pm each day, asking was there was anything to report. This would be sent to Holyhead and initially, a single word answer would be sent back; either 'Yes' or 'No'. In other words, the message and reply would travel a total of 144 miles. According to the newspaper, if this took a minute, it was regarded as being very slow work on the part of the semaphore operators. A quick calculation tells us that if the message and reply took one minute to travel 144 miles, then the overall speed at which the information was being relayed was 8,640mph. In practice, the actual speed achieved was almost invariably far greater than this. The journalist writing the piece from the *Shrewsbury Chronicle* looked at the time taken in September 1836 and found that the five fastest questions and answers that month had taken a total of two minutes and 36 seconds. Since the distance travelled by the messages in that time had been 720 miles, this worked out at 288 miles per minute or 17,280mph. This is approximately the speed at which the International Space Station is orbiting the Earth.

The speeds cited above are truly extraordinary for an arrangement of purely mechanical devices, but the fastest time of all for the enquiry to be sent from Liverpool and the reply received from Holyhead is simply staggering. On 10 September, it took only 25 seconds for the message to reach Holyhead and the reply to be received. This equates to a speed of over 20,000mph. A message sent from London to Moscow at this speed would take less than five minutes to arrive.

It is easy to see why the optical telegraph was thought to be perfectly satisfactory for all practical purposes and why, even when the electric telegraph began to be developed, there were those who were quite happy to stick with semaphores and shutters. True, there were problems in bad weather and it was impossible to send any messages at all in the hours of darkness, but the methods were reliable and extremely fast. By contrast, when the first telegraph wires were rigged up, they were prone to failure and very vulnerable to sabotage. A single wire which went across country for hundreds of miles seemed a very fragile and uncertain way of communicating. The slightest accident, let alone deliberate acts of sabotage or vandalism, would be enough to put it out of action.

Not only did electric telegraphs, when they began to appear, seem very feeble and delicate things when compared to the stout and highly visible optical telegraphs, there was widespread scepticism about them to begin with. Anybody could understand, simply by watching them in action, how the existing telegraph lines worked. Not so the electric lines, which caused a needle to twitch weakly or made a succession of buzzes. To begin with, the whole thing looked like some kind of scam and it required some very public demonstrations to persuade the average person that communication by electric wire was not just some elaborate confidence trick.

It was the development of electric telegraphs in the first half of the nineteenth century which spelled the end for optical telegraphy. In England, the installation of electric telegraph lines alongside railway lines coincided precisely with the dawn of the Victorian era, with Victoria's accession to the throne in 1838. Six years later, Samuel Morse demonstrated the capabilities of long-distance communication by this method, when he sent the famous message 'WHAT HATH GOD WROUGHT?' from Washington to Baltimore in 1844.

In retrospect, the advantages of the electric, as opposed to the optical, telegraph are so blindingly obvious that it is a wonder that not everybody at the time saw them. Nevertheless, there was strong opposition to the electrical telegraph at first and many commercial enterprises preferred to stick with the familiar and more easily understood technology of mechanical arms and, on occasion, steam power. At the same time that the Great Western Railway was arranging for the laying of an electric telegraph line running 13 miles between West Drayton Station and the terminus at Paddington, another railway company was rejecting electricity in favour of a pneumatic system which signalled with whistles. Ten years after the electric telegraph was linked up to London's Paddington Station, the British Army wished to set up a telegraph network on the Mediterranean island of Malta. Surprisingly, they opted for a purely mechanical system and semaphore towers were built across the island.

The eventual triumph of electrical telegraphs over semaphore towers took some decades and it was economic considerations which proved decisive in bringing about the end of the embryonic mechanical Internet. Semaphore lines were very costly, both in terms of the physical structures needed and also in manpower. When Samuel Morse sent his message from Washington to Baltimore, only two people needed to be involved; one sending the words in Washington and another to receive and decode them 44 miles away in Baltimore. The average semaphore line covering such a distance would have needed six or seven stations, at intervals of about eight miles each, and each station would need a staff of at least three people. In other words, to send the simple message, 'What hath God wrought?' would have needed perhaps twenty people, rather than just two. It was calculations of this kind which led inexorably to the decline and fall of optical telegraph networks.

Meanwhile, in the early years of the electric telegraph, optical telegraphs continued to forge ahead and to many people, they still looked like the future of telecommunications. In Russia, at the same time that Britain and America were tentatively experimenting with electric wires, a new line was completed in 1839 which connected St Petersburg with Warsaw. As late as 1854 a completely new design of optical telegraph was used for the construction of a line of eighty stations running along the southern coast of Finland. This ran from Helsinki to Turku and also into Russia.

Even when the electric telegraph had proved itself and was flourishing across the world, there was a reluctance to do away with the older telegraph stations. The Liverpool to Holyhead line, for example, continued operating until 1860. The last regular use of a European semaphore telegraph was in Sweden, which used one to communicate between the mainland and an offshore island until 1880.

The extensive lines of semaphores and shutters described in steampunk fiction have been shown to be no work of the imagination but rather a very real feature of the eighteenth and nineteenth centuries. Something about mechanical telegraphs seems to capture the imagination. They are to be found not only in steampunk novels, but also in stories of alternative history. A novel from 1939, *Lest Darkness Fall* by L Sprague de Camp, tells the story of a twentieth-century American who is mysteriously transported back to the days of ancient Rome. The plot device is reminiscent of Mark Twain's *A Connecticut Yankee at the Court of King Arthur*. The protagonist of *Lest Darkness Fall* determines to avert the Dark Ages which occurred after the collapse of the Roman Empire in the sixth century AD. He introduces the Roman world to things such as printing and bookkeeping. To maintain communications, the hero arranges for the construction of a semaphore chain.

The optical telegraph still crops up in modern science fiction. In the *Safehold* novels of David Weber, humanity is all but destroyed by an alien civilization. The scattered remnants find refuge on a distant planet. Their only hope is to live quietly, without attracting the attention of the enemy by displaying any signs of technology or industrialization. In order that radio waves do not leak out into space and thus betray their presence to the enemy, the colonists rely upon semaphore towers to communicate between towns. *Terminal World* by Alastair Reynolds is a story set in the distant future, which tells of the last remaining great city. This is connected to outlying settlements by ancient semaphore stations.

There exists in Britain a lingering relic of the mechanical Internet at which we have looked in this chapter. In the early 1840s, when optical telegraphs were at their height in Europe, an engineer called Charles Hutton Gregory wondered if it might be possible to adapt the semaphore telegraph and use it as a reliable form of indicating to railway trains when to stop and when to go, as well as a few other simple instructions and warnings. There was a variety of different ways at that time of signalling to trains, but no one accepted system. Gregory obtained permission to install semaphore signals on the London and Croydon Railway and they proved a great success.

In essence, semaphore signals as used on the railways were like a form of traffic light. When the arm was in one position, this meant go. In another and the engine driver knew that he should stop. Unlike the semaphore telegraphs, railway semaphores worked as easily at night. Oil lamps were lit and different coloured lenses moved to expose or obscure the appropriate light. For well over a century, semaphore signals were a regular feature of British railway lines and also those of many other countries. Since the end of the Second World War though, they have gradually vanished; replaced by more up-to-date electrical signs. A few remain, but they too will soon be replaced. Within a decade or so, the last visible sign of semaphore telegraphy will vanish.

Chapter 4

Steam Planes Take Off

It is one of the few things that we all know about the history of flight; that the Wright brothers were the first to take to the sky in a heavier-than-air flying machine. They achieved this feat in 1903, a couple of years after the death of Queen Victoria. Aeroplanes were unknown before the twentieth century. This knowledge makes the appearance of powered flight in steampunk narratives set in the nineteenth century pleasingly anachronistic. When we read a description of steam-powered aeroplanes, we know at once that we are firmly and definitely in the realm of fantasy. We looked in Chapter 1 at the transatlantic vacuum railway which appears in *A Transatlantic Tunnel, Hurrah!* While the tunnel of the title, more properly a submarine tube, is being constructed on the seabed, we are given glimpses of the other methods of getting around the world in the alternative universe posited by Harry Harrison, the author. One of these are the coal-fired aeroplanes which soar overhead.

The aeroplanes in *A Transatlantic Tunnel, Hurrah!*, although fuelled by coal, are not steam planes. They use a quite different means of propulsion, one which shows once again the way in which the authors of steampunk often have a remarkable knowledge of obscure historical facts. Because, strange to relate, coal-fired aeroplanes were being developed by Nazi Germany towards the end of the Second World War and there are even rumours that a jet plane fuelled by coal was the first ever to break the sound barrier.

In his steampunk novel, Harry Harrison describes in great detail an enormous and luxurious flying boat which carries passengers across the Atlantic; between Southampton and New York. The *Queen Elizabeth* is a turbojet, using powdered coal as a fuel. This is perfectly feasible and powdered coal was used as a fuel for a number of strange engines in the nineteenth century. The idea of using it to power a transatlantic flight though sounds like a sheer and exuberant flight of Harry Harrison's famous imagination. In fact, he just read up on the history of flight.

After America's declaration of war on Germany in 1941, thought was given in Germany to a way of carrying the war to the American mainland, by means of a long-range bomber. Many ideas were mooted for the so-called

'Amerika' bombers; one of which must surely have provided the pattern for the transatlantic airliner in *A Transatlantic Tunnel, Hurrah!*.

One of the vital assets when waging war in the developed world is a ready supply of oil for one's aeroplanes and tanks. Germany of course, like Britain, relied on importing this precious commodity, because they possessed no subterranean deposits of their own. Some of the fighting during the Second World War centred around seizing and occupying territory which had oil. Germany and Britain might not have had much oil of their own, but they both had great amounts of coal and this was mooted by the Germans as a means of providing fuel for a bomber capable of making the 7,200-mile round trip to the United States. A test-bed was set up for a 6,000hp turbine which was to operate on pulverized coal dust. Unfortunately, it was destroyed in an air raid before it could be fully tested.

Another coal-fired aircraft was to be the Lippisch P.13a fighter, able to reach Mach 3 or three times the speed of sound. The P.13a was designed as a ramjet, with a basket of coal at the front of the jet, which would be ignited by a jet of gas. In Harry Harrison's plane, the coal dust was to be ignited in precisely the same way, by a stream of burning butane. There are rumours that a prototype of the Lippisch P.13a was actually built and tested. According to these unsubstantiated stories, the delta-winged fighter broke the sound barrier a few years before American pilot Chuck Yeager achieved this feat officially in 1947.

Anybody reading Harry Harrison's novel of an alternative universe might very well suppose that he had dreamed up a fantastic and wholly novel means of powering an aeroplane, whereas the reality is that he had done his homework and needed only to write of things which had taken place in the real world a few decades earlier.

Before we look at the idea of actual steam-powered aircraft in the nineteenth century, let us return to the question of the Wright brothers and their famous flight in 1903. What did they actually achieve and for what are we remembering them? Not the first manned flight of course; that took place in France in 1783, when various people ascended in hot air and hydrogen balloons. The first heavier-than-air flight? That honour goes to a 10-year-old boy who, in 1849, flew in a primitive glider devised by Englishman George Cayley. A few years later, an adult flew in another of Cayley's gliders, this time one with controls which the pilot operated in order to manoeuvre the aircraft in flight. Was it perhaps powered flight that the Wrights first demonstrated? Not even that, for as early as 1853, Frenchman Henri Gifford used a steam engine to propel a dirigible through the air above Paris. Powered heavier-than-air flight? The Wright brothers were beaten to that goal too by others, including a Frenchman and a fellow American. We shall return later in the chapter to the question of what it actually was that Orville and Wilbur Wright are remembered for. Readers will also learn that the Wright brothers' place in history is ensured by a secret document which discourages any investigation into their claim to have piloted the world's first aeroplane and that the facts are a good deal more complicated than most of us realize.

There may be controversy about the first successful aeroplane flight, but there is none at all about the world's first airline company. In 1843, just five years after Victoria came to the throne, the Aerial Transit Company was incorporated in London. According to publicity material distributed by the new company, the object was to build 'aerial steam carriages' which would 'convey passengers, troops and government dispatches to China and India'. Illustration 3 shows one of the aeroplanes which would make this ambitious project a practical proposition. Throwing out smoke, like an airborne railway engine, the airliner sails over the heads of admiring crowds. It very much represented the spirit of the times. It was only a few years since the first fare-paying passengers had been transported by a railway train and railway mania was in full flow. If steam engines could carry people across the sea and from city to city, why not into the sky as well?

The very idea of a steam plane seems today so patently absurd that we smile involuntarily at the thought. Who could imagine such a thing, outside the pages of a fantasy? In fact, plenty of people not only imagined steam aeroplanes; they also built them. And while it is perfectly true that some were ludicrously unsuited to flying, others did indeed leave the ground, quite literally under their own steam. Nobody who studies the matter for more than an hour or two doubts for a moment that the first heavier-than-air flying machines were powered not by primitive internal combustion engines but rather by sophisticated steam engines which had been many years in development. The real mystery is why this is not generally known.

The principles of flight were understood a hundred years before the Wright brothers began their experiments. Around 1800 a well-to-do young man called George Cayley was experimenting with flying machines. At that time, it was generally thought that birds flew by flapping their wings, but Cayley was interested in the idea of wings being used instead for gliding. The ornithopter, a flying machine with wings which flapped like a bird, was a blind alley as far as manned flight was concerned, although a great deal of effort was expended in the nineteenth century on trying to build such devices. After long and hard thought, combined with a good many practical experiments, Cayley decided that a cambered surface would be the best shape to support a vehicle in the air. He constructed something very similar to a wind tunnel and in 1809 wrote up the results of his researches in a paper called 'On Aerial Navigation', which was published in *The Journal of Natural Philosophy, Chemistry and the Arts*. George Cayley had many other interests and having analysed some of the problems associated with flying, he turned his attention to other matters for the next 35 years or so. George Cayley is sometimes referred to, with good reason, as the 'Father of Aviation', but his work had an influence in other areas too. In 1808, he designed a tension-spoked wheel which he thought would be effective for the undercarriage of an aircraft. Today, we see this kind of wheel every day on bicycles.

Other people, apart from George Cayley, were also intrigued by the idea of flight by means of aircraft with fixed wings. Although we take it for granted today that the wings of aircraft should be both rigid and also secured firmly to the body of the plane, this was a radical notion 200 years ago. Most inventors were thinking in terms of flying by flapping wings as birds do. Two men who shared George Cayley's unorthodox views on this subject were John Stringfellow and Samuel Henson. In 1841 Henson patented a lightweight steam engine, which he thought powerful enough for an aeroplane. The following year, he teamed up with John Stringfellow, an engineer, and the two of them designed an aeroplane with a 150ft wingspan, which they thought would be capable of carrying not only a pilot, but also passengers and cargo. In March 1843, they were granted a patent for the world's first powered aircraft. The technical drawings for this monoplane show something which looks very much as though it might be able to take to the skies if only a powerful-enough source of energy could be found to keep the propellers turning fast enough. It has cambered wings, moveable control surfaces and all the features that we might expect to see on a modern aeroplane. An artist's impression of this invention may be seen in Illustration 6, where it is depicted soaring above London.

Samuel Henson and John Stringfellow built various models of their giant airliner, fitting them with very compact and efficient steam engines to drive the twin propellers which were supposed to provide the motive force to lift the aeroplanes from the ground. These experiments continued for several years and culminated in the construction of a huge model with a 20ft wingspan. Like all the previous models, this too suffered from one great and insurmountable problem; it could not fly. The things worked well enough as gliders; they were all put together on sound aerodynamic principles. The difficulty lay in the engines, none of which were powerful enough to propel the planes at a great enough speed for them to achieve lift-off.

It was all very well applying for a patent for an aeroplane, but Henson went a little further than this, having brochures and company prospectuses printed in which an airline dependent on the as yet untested aeroplane was presented as being just around the corner. Of course, this was an exciting time for steam-powered travel and a railway boom was in full swing. It was not inconceivable that steam travel by air might be on the horizon. The Aerial Transit Company was incorporated in 1843 to raise money so that a full-sized aeroplane could actually be built, something which had so far eluded Henson and Stringfellow.

The coloured drawings prepared in the 1840s to publicize Henson's aerial steam carriages look like illustrations for the cover of a steampunk novel. The giant aircraft is shown above London, sailing over the pyramids of Egypt and landing in India. These vivid images certainly captured the public imagination. There were so many exciting new inventions appearing at that time, that it did not at first seem incredible to the Victorians that they might soon be taking to the sky by means of the same motive force which enabled them to travel along railway lines. It was only as the years passed and no actual flying machine appeared that doubts began to creep in. Fraudulent companies were not uncommon in nineteenth-century Britain and it gradually began to seem that Henson's airline company was one of them. Excitement turned to ridicule and Samuel Henson became disillusioned with the idea of a heavier-than-air flying machine and left the country for America, where he invented something a good deal more practical than a steam-powered airliner. Henson's T-shaped safety razor set the shape for all the razors that we use today.

In Chapter 1, we saw how very nearly self-powered road vehicles became the dominant means of transport in Victorian Britain and the difference that this might have made to society at that time. What on earth would the British Empire have been like if fleets of steam planes were at London's disposal, to carry troops anywhere that the queen required? This really has the makings of a steampunk novel; steam-powered airliners linking the various parts of the empire.

According to the wildly optimistic publicity material being turned out by the Aerial Transit Company, these aeroplanes had a military as well as civil potential. It would supposedly be possible to transport troops from one part of the empire to another at great speed. The aircraft would be able too, of course to carry mail, freight and passengers.

John Stringfellow did not give up so readily on the idea of powered flight. After the departure of Samuel Henson, Stringfellow, assisted by his son Frederick, continued to experiment with model aeroplanes and in the summer of 1848 had a breakthrough; the world's first powered, heavier-than-air flight. The Stringfellows' family business, based in the Somerset town of Chard, was lacemaking. A long gallery on the upper floor of their factory was chosen as the place to test their latest aeroplane. There were two reasons for undertaking the test indoors. One was that wind blowing in the wrong direction had sabotaged previous attempts. Aeroplanes take off because of the flow of air over their wings; from front to back. If this is disrupted, then insufficient lift may be generated for them to leave the ground. This was particularly a problem in those early days, when the principles of aerodynamics were still only hazily understood.

The other reason that the long indoor gallery was the perfect place to test the aeroplane was that it was possible to fix up a guide wire, allowing the plane to accelerate with hardly any friction to impede it, until it was moving fast enough for the wings to lift it. Gaining sufficient speed in this way prevented many early aeroplanes from taking off. There is a lot of friction involved when an aeroplane trundles along on wheels and for many early models, this prevented them from moving fast enough for their wings to generate the lift necessary for them to fly. Even the Wright brothers encountered this problem. Their first aeroplane, the *Flyer*, did not have any undercarriage at all. Instead, the Wrights laid out a miniature railway track and their plane lay on a trolley running smoothly along this, from which it rose when it was moving fast enough. Without this assistance, combined with a strong headwind, that famous first flight which the Wright brothers undertook in 1903 would never have been able to take place.

The aeroplane, which carried out what was indisputably the world's first successful flight by a powered, heavier-than-air machine, used a very small steam engine weighing a little over 6lbs, including water and fuel. Heat was provided not by coal, but by a compact oil lamp. The two contrarotating propellers sent the plane rushing along the guide-wire until it lifted away from the wire and flew straight and true for over 30ft. It was a fantastic victory for Stringfellow's theories and he later went on to make a large model triplane which he demonstrated in front of the Prince of Wales in the Crystal Palace. There could be not the least doubt any more that powered, heavier-than-air flight was possible. The only question was not if, but when, a manned flight would be made.

George Cayley, who had laid the foundations for a lot of what had been going on in the field of aerodynamics during the first half of the nineteenth century, had not taken an active role in things since the publication of his paper on the subject in 1809. Samuel Henson had tried unsuccessfully to draw Cayley into his plans for an airline, but had had no luck. Perhaps the letters from Henson had reignited Sir George's interest in manned flight, because during the 1840s, he built and tested at least two gliders. One of these, a biplane, took off in 1853 with a pilot in it.

In 1851, the 15 September issue of *The Mechanic's Magazine* carried an article by Cayley, describing plans for a steerable, unpowered aeroplane which he hoped to build. This was to have a single wing, 500ft in area, and would weigh around 300lbs. The pilot would sit in a compartment like a small boat, fitted with a tricycle undercarriage, and a tiller would control a rudder, enabling a change of direction if required.

The machine described in *The Mechanic's Magazine* was constructed to George Cayley's specifications and tested in the summer of 1853. Cayley's home was in Yorkshire and the place chosen for what would be the world's first flight of an adult in a controllable, heavier-than-air machine was a small valley near Brompton Hall, Cayley's home, near Scarborough. The pilot may not have been altogether a volunteer; he was Sir George's coachman. There were a number of eyewitnesses to the event, including George Cayley's granddaughter, who was 10 at the time. In 1921, she wrote a letter containing a detailed account of what took place that day. She said;

> Everyone was out on the high east side and I saw the start from close to. The coachman went in the machine and landed on the west side at about the same level. The coachman got himself clear, and when the watchers had got across, he shouted, 'Please, Sir George, I wish to give notice. I was hired to drive, not to fly.'

The distance of this flight was estimated at the time to be about 500 yards.

Before going on to look at the first flights of powered aeroplanes, steam planes, we shall consider another of the recurring themes in steampunk; steerable balloons or dirigibles. Some people in the nineteenth century thought that airships were the route which would lead to the conquest of the skies, rather than the clumsy-looking aeroplanes being developed at that time. Airships looked far more graceful and were felt to be a more practical option for travelling by air.

Steerable balloons have featured in steampunk novels from the very beginning. One of the earliest of the proto-steampunk novels is *Bring the Jubilee* by Ward Moore, first published in 1955. It tells the story of life in the United States from 1938 until the early 1950s. This is not the world that we recognize though. The point of divergence was the American Civil War, which in *Bring the Jubilee* was won by the South. This has left the northern states impoverished and technologically backward, which has in turn had a knock-on effect upon the whole world. There are no telephones, for example. Instead, there are private telegraphs in people's homes. The only cars are 'minibiles', driven not by internal combustion engines but steam. Heavier-than-air flight is regarded as a chimera and only airships are seen overhead.

Another proto-steampunk novel, Michael Moorcock's *The Warlord of the Air*, 1971, shows the world of 1973 as it might be if aeroplanes had not been invented. The empires of the world, chief among them that of the British, maintain their hold on their colonial possessions by means of fleets of airships. More recent steampunk novels, such as those of Philip Pullman and Philip Reeve, also feature balloons as a mode of travel in alternative versions of British cities.

Readers might perhaps be surprised to learn of the varied roles that balloons played in the nineteenth century. Their use in time of war led to the development of both the first aircraft carrier, in 1861, and also the first anti-aircraft gun, in 1870. Both ideas sound weirdly out of place. Who can imagine an aircraft carrier being used in the American Civil War?

The first balloons to carry men and women off the ground took flight in France in 1783. These were hot air balloons and very dangerous they were too. A prolonged flight entailed lighting a bonfire in close proximity to an inflammable balloon made of paper or fabric. Shortly after the first flights in hot air balloons, the gas hydrogen began to be used. This was more convenient and some people saw at once that balloons filled with hydrogen might make it possible to travel by air. The problem was that balloons were wholly at the mercy of the wind. If they were to be used for anything other than the occasional joyride, then some way of controlling them and making it possible for them to fly against the wind would have to be found.

One problem with balloons is that if they are spherical, then there is no front or back and steering them is likely to prove tricky. The only way that a practical airship could be developed would be to have a long, cigar-shaped balloon, which could be aimed and propelled in a certain direction. At the same time that Samuel Henson was trying to raise money for his airline, a rival company emerged in the field, which also promised to be able to arrange international travel by air. Posters appeared in London, under the heading European Aeronautical Society, which described a new airship called the *Eagle*. This was for the purpose of 'establishing direct communication between the several capitals of Europe'. A trip from London to Paris and back again was announced. The new airship was said, in the posters, to be 100ft long and 50ft high and to boast a crew of seventeen. There wasn't a word of truth in it and the *Eagle* was no more capable of flying to Europe than Henson's aerial steam carriage was of getting to India.

The enterprise advertised in London was a little ambitious, but one of the men involved in it, Henri Giffard, went on to achieve another world first in flying; the first powered and controlled flight. It is hard to imagine who would think it a good idea to combine a coal-burning steam engine with a huge balloon filled with a highly inflammable mixture of hydrogen and methane gas, but in 24 September 1852, that is exactly what Henri Giffard did. Using a steam engine weighing 350lbs, he propelled a dirigible for 17 miles, at a speed of roughly 6mph. The airship was, for the first time, shown to be a practical means of travelling from one place to another. Giffard's airship may be seen in Illustration 14.

Less than a decade after Giffard's flight across Paris, balloons were harnessed for military purposes. The advantage of being able to spy on an enemy's positions by rising high in the sky are obvious. When the American Civil War began in 1861, the Federal army of the North quickly set up an airborne corps. The American Army Balloon Corps had a strength of fifty men and seven balloons at its disposal and it was commanded by a man called Thaddeus Lowe. On 18 June 1861, Lowe achieved another first for aviation, when he installed a telegraph transmitter in one of the balloons and sent a message to President Lincoln, while airborne. He said;

> I have pleasure in sending you this first despatch ever telegraphed from an aerial station, and in acknowledging indebtedness for your encouragement for the opportunity of demonstrating the availability of the science of aeronautics in the military service of the country.

The army balloons proved so successful for observations that a method was devised of moving them swiftly from place to place. The gas which filled the balloons, hydrogen, could be manufactured by pouring sulphuric acid over iron filings. Rather than carry sacks of iron filings about in horse-drawn carts, somebody came up with the idea of moving both them and the deflated balloons about along rivers and canals. A coal boat called the *G.W. Parke Custis* was rebuilt as a mobile launching platform for observation balloons. The ingredients necessary to generate the hydrogen could be moved about far more easily than overland. This boat became, in effect, the world's first aircraft carrier.

Five years after the end of the American Civil War, a war in Europe saw another innovation in matters relating to aerial warfare, when the German munitions manufacturer Krupps designed and made the world's first anti-aircraft gun. For reasons which need not concern us here, France and Prussia, the main part of modern-day Germany, went to war in 1870. The Prussians laid siege to Paris and the city was cut off from all communication with the outside world. On 23 September 1870, a balloon was launched from Paris, carrying 224lbs of letters. This became a regular service; the first airmail service in the world, in fact. The Prussian troops fired their rifles at the balloons, but they were too high to hit. The arms company Krupps was commissioned to make an artillery piece which fired vertically upwards. The *Ballonabwehrkanone*, or anti-balloon cannon, brought down five balloons.

In the steampunk series of *Mortal Engines* books by Philip Reeve, hot air balloons are used extensively, both for travelling from one part of a city to other, a little like taxis, and also to visit other cities. This idea, of balloons as buses or taxis, was an enticing one in the nineteenth century, although it never led anywhere. Steam engines were heavy and dangerous and so another means of driving the propellers on airships was tried. France was the country most associated with airships and balloons and it was there, on 9 August 1884, that two army engineers, Charles Renard and Arthur Krebs, took a gigantic new airship on its maiden voyage. *La France* was 170ft long and powered by an electric motor which delivered over 8hp. This allowed *La France* to travel against the wind, averaging 13mph.

A twentieth-century world dominated by airships and balloons, a staple of steampunk fiction from the 1950s onwards, seems to us grotesque, but only because we are so familiar today with aeroplanes that we suppose their advantages for transport and all other purposes to be unassailable. *Bring the Jubilee* set the trend for this version of the modern world and it has been eagerly seized upon by other writers, who feel that showing readers a sky filled with dirigibles is a sure-fire way of setting the scene for a steampunk extravaganza. Warde Moore's description of an alternative New York in 1938 is worth quoting at length, for it has become the template for many similar stories. In this book, the United States lost the American Civil War and while the Confederacy has become a world power, the United States is a backward and primarily rural nation. After describing the amazing sight of buildings twelve or fifteen storeys high, the protagonist looks beyond these tall buildings to the sky overhead;

> Above them balloons moved gracefully through the air, guided and controlled as skilfully as old-time sailing vessels. These were not

entirely novel to me; I had seen more of them than I had minibiles, but never so many as here. In a single hour, gawking upwards, I counted seven, admiring how nicely calculated their courses were, for they seldom came so low as to endanger lives beneath by having to throw out sandbags in order to rise. That they could so manoeuvre over buildings of greatly uneven heights showed this to be the air age indeed.

In Michael Moorcock's *The Warlord of the Air*, like *Bring the Jubilee* a proto-steampunk novel, airships dominate the world and are a key plot device. More recently, S.M. Stirling used the same idea to great effect in *The Peshawar Lancers*, first published in 2002. Like the lack of petrol-driven cars, the presence of airships is a handy motif for setting the scene of an alternative universe.

We must of course bear in mind that even in our own world, aeroplanes were at first regarded as being an irrelevant novelty, at least as far as carrying passengers or being of any military use were concerned. Ten years after they had become well-known through the publicised achievements of the Wright brothers, a lot of people thought that heavier-than-air flying machines would never amount to anything. For such people, the mighty airship was the way forward, rather than the flimsy little aeroplanes which struggled to carry more than one person at a time. Before returning to the potential of powered dirigibles, let's just see some of the discouraging remarks being made about the possibility of aeroplanes being capable of anything other than joyrides.

In 1910, a year after Bleriot had crossed the English Channel in an aeroplane and heavier-than-air flight had ceased to be a mere curiosity, eminent Harvard astronomer William Pickering had this to say about the prospect of aeroplane flight;

> The popular mind often pictures gigantic flying machines speeding across the Atlantic, carrying innumerable passengers. It seems safe to say that such ideas must be wholly visionary. Even if such a machine could get across with one or two passengers, it would be prohibitive to any but the capitalist who could own his own yacht.

Military men, who might have been expected to be a little more open to the idea of a new machine which could prove useful on the battlefield, were no more enthusiastic. The Wright brothers offered the British Admiralty an option on their patents, but the response they received was dismissive. In 1907, the

Admiralty said of powered aeroplanes, 'Their Lordships are of the opinion that they would not be of any practical use to the Naval Service.' The British Army was no more interested than was the navy. Richard Haldane, Secretary of State for War, said in 1910 that, 'We do not consider that aeroplanes will be of any possible use for war purposes.' The French couldn't see any point in aeroplanes either. Marshal Foch, the brilliant French strategist who would be Commander-in-Chief of the Allied armies fighting against Germany in 1918, gave it as his considered view in 1911, that 'Airplanes are interesting toys but of no military value'.

Those who rejected the aeroplane, even when long-distance flights had been demonstrated, placed all their hopes for the future of air travel and the military exploitation of the skies, in airships. This attitude was still going strong as late as the 1930s, with enormous dirigibles such as the British *R101* and the German *Hindenburg* offering passengers the luxury of an ocean-going liner which sailed above the clouds.

European armies had of course adopted balloons and seen their use for observation, long before the invention of the aeroplane. The British, for instance, were using them in colonial wars in Africa as early as 1884. By 1907, they had commissioned the first dirigible and perhaps it was felt that the pace of change was growing a little frantic. Just as they launched their first manoeuvrable airship, along comes an American hoping to sell them the design for some new-fangled, heavier-than-air machine that would make their balloons obsolete!

The point being made is that there was no certainty about the ascendency of the aeroplane, if the pun will be forgiven. In Germany, airships were the focus of interest as far as flight was concerned. One scheme in that country could have been lifted straight from the pages of a steampunk narrative and that was the development of a metal airship. The very idea sounds so preposterous that it might be of interest to look at this strange enterprise.

The first design for an airship was drawn up in the late seventeenth century, when a Jesuit priest called Francesco de Lana published in 1670 a booklet called *Demonstration of the Feasibility of Constructing a Ship With Rudder and Sails, Which Will Sail Through the Air*. The idea was sound, if impractical. The proposal was that if four very large copper spheres could have all the air pumped out of them, then the buoyancy of the air would cause them to float upwards. If they were big enough, then surely a boat could be attached to them and also be lifted into the air? If such a thing had been attempted, then the metal spheres would have been more likely to have been crushed by the pressure of the air than anything else, but this was the first serious attempt to design an airship.

In 1844, a determined effort was actually made to construct a metal balloon. A large workshop was established in a Paris back-street by a man called Edmond Marey-Monge. With the help of a team of workmen, he painstakingly soldered together countless strips of very thin brass, which were to form the outer layer of what was known as the *ballon de cuivre*, the brass balloon. When completed, Marey-Monge hoped to fill it with hydrogen gas.

The rationale behind Marey-Monge's endeavour was sound. Writing in the magazine *Comptes Rendus*, he pointed out quite correctly that balloons at that time were fragile, being made of cloth, paper and wood. If there was to be a realistic possibly of aerial navigation turning into a regular activity, then the airships must be in a 'position, like our ships, to resist the bad weather of ten or fifteen years' service'. This meant building them of a durable substance, and what could be more durable than metal?

In a cavernous shed, at the end of an alleyway behind the newly-built Montparnasse railway station, the great balloon took shape. Visitors paid to be admitted to the workshop to marvel at the progress of this latest product of the Industrial Revolution. So many exciting new inventions were appearing in Europe at this time, that there was no reason to suppose that air travel by a gleaming brass balloon would not soon become the next big thing.

The chief difficulty was of course that metal weighs a good deal more than fabric and glue. The sheets of brass, each 15ft long, were, admittedly very thin. So thin that no factory or foundry in France was able to work to such specifications; they were imported from Prussia. Even so, the 30ft-wide balloon would still weigh almost 900lbs. This was without the gondola or passengers. As the brass panels were soldered together, it was obvious that small gaps existed and so Marey-Monge coated the inside of his balloon with paper and glue.

The problem is that hydrogen atoms are very small indeed and will escape from even the tiniest hole, as the inventor discovered when the great brass sphere was completed and it was time for the test. It was a failure, leaking dangerously in a dozen places, and never took off. As fast as the hydrogen gas was pumped in, it escaped and so the first test, which took place on 2 June 1844, was also the last. The balloon was subsequently dismantled and sold for scrap.

The idea of a metal airship was an intriguing one and others tried to build such a craft. The advantages for a dirigible of being made of metal were considerable. The wrinkly skin of the average airship has the opposite effect to streamlining, creating drag and slowing the thing down to a crawl. If the surface were to be rigid, hard and smooth, then very much greater speeds could, in theory, be attained. Patents were applied for over the next half century, including one for a 400ft-long steel airship. It was to be half a century after the failure of the brass balloon that the flight of the world's first metal airship took place.

David Schwarz was a Hungarian inventor who claimed to various European governments that if they would only fund him, he could construct for them a revolutionary aircraft; a rigid, metal airship which would need no balloons to hold the gas which lifted it. The Austro-Hungarians were at first interested, but then declined to become involved. The Russian military attaché in Budapest agreed in 1893 to support Schwarz and he was given facilities to travel to Russia and build what the Russians hoped would be a world-class weapon of war. After considerable work and many delays, it became apparent that the dirigible was leaky and could never take off. Schwarz left Russia hurriedly, under mysterious circumstances.

Rumours began to circulate that David Schwarz was either a madman or a confidence trickster, but this did not prevent a businessman teaming up with him and obtaining a contract with the Prussian government to build the metal airship. It was clear that it was the military applications of the thing which really interested backers, because the Prussian Airship Battalion was ordered to cooperate in the enterprise and facilities were provided at Tempelhof, near Berlin, later to be the site of Berlin's airport. Work went ahead steadily until 13 January 1897, when Schwarz suddenly and unexpectedly dropped dead from a heart attack.

Despite the death of the man who had invented it, work continued on the airship to which he had devoted so much time and effort. The entire machine was made of aluminium, for lightness. The outer shell was of aluminium plates which were just a fraction of a millimetre thick and even the propellers, driven by a Daimler engine were made of the same metal. On 3 November 1897, it was time for the maiden flight of what was perhaps the most extraordinary flying machine of the nineteenth century.

A soldier from the Airship Battalion was chosen as the pilot and once the dirigible had been filled with hydrogen, it shot up into the air with surprising velocity. Then things started to go wrong. First one propeller and then another stopped working and the ungainly airship ended up side-on to the wind. Fearing for his life, the pilot, Ernst Jägels, opened the gas release valve and the great airship descended rapidly from 500ft to the ground, where it crumpled up like tinfoil. There never was a second flight.

We view today with amusement the efforts to build a metal airship, but the idea is not quite as fanciful as it might seem at first sight. The difficulty was not with the actual concept, but more that the necessary technology for such a project did not exist in Victorian Europe. In the 1920s, when airships were enjoying something of a vogue, the United States Navy commissioned the Detroit Aircraft Corporation to construct a 150ft-long airship which would be made entirely of metal. The *ZMC-2*, which stands for Zeppelin Metal Clad 200,000ft^3 capacity, required the invention of new techniques for riveting together sheets of aluminium. Helium was to be the lifting force, rather than highly-inflammable hydrogen and when the airship was finished, it was found that it could travel at 70mph, an almost unbelievable speed for a dirigible. This was due to the fact that the skin was quite rigid and entirely smooth.

For a while, there were plans to build an even larger metal airship, one which would have rivalled the German *Graf Zeppelin*. However, the Depression arrived and there were other uses to which the money was put. The *ZMC-2*, known affectionately in the navy as the 'Tin Bubble', continued in service until 1941. There were tentative plans to use it for anti-submarine warfare, but eventually it was scrapped.

Metal airships were a perfectly practical proposition and had it not been for the terrible disasters of the *R101* and the *Hindenburg*, it is altogether possible that the twentieth century would have been remembered as the time when airships came into their own, rather than being dominated, as it was, by heavier-than-air flying machines. For steampunk, the replacement of aeroplanes by dirigibles is a handy, shorthand way of showing a more backward world than the one which we know.

It was mentioned earlier that airships and balloons are a common feature of steampunk narratives, but of course they are found too in the works of writer such as Jules Verne. Both *The Mysterious Island* and *Five Weeks in a Balloon* by Verne have manned balloons as a central plot device. Interestingly, the one book by Jules Verne which everybody associates with balloons, *Around the World in Eighty Days*, does not feature any travelling by balloon. Balloons in the nineteenth century were exciting novelties which found their way into other speculative fiction. Perhaps this was because some very startling things were being achieved by balloonists at that time. They were, for one thing, exploring the stratosphere.

It may seem an amazing idea that Victorian aeronauts were travelling as high as modern supersonic jets fly; it is nevertheless true. Two of these intrepid explorers, whose adventures quite possibly inspired the writers of late-nineteenth century fantastic fiction, were based at Wolverhampton gasworks, in the heart of England's industrial midlands. James Glaisher, who had for over 20 years been the superintendent of Greenwich Observatory's

Magnetical and Meteorological Department, teamed up with famous balloonist Henry Coxwell to see how high above the earth it was possible to make scientific observations.

James Glaisher was a member of the British Association for the Advancement of Science, which had for years been trying to arrange for a high-altitude balloon flight to gather data about the upper atmosphere. In 1861, Henry Coxwell offered to build a balloon which would be big enough for this task, but only on the condition that the British Association for the Advancement of Science would then hire it for their research at £50 a flight. They agreed and so Coxwell built a huge balloon, which he named *Mammoth*. This had a volume of 93,000ft^3. When the *Mammoth* took off for its first flight, on 17 July 1862, neither of the two men in the basket slung beneath it were youngsters. Glaisher was 53 and Coxwell 10 years younger.

The ascent was made from the Wolverhampton gasworks for the simple reason that almost 100,000ft^3 of coal gas were required to fill it to capacity and the gasworks was the best place to obtain that. Various instruments for measuring the altitude, temperature, air pressure and so on accompanied the two men when they headed up into the sky, including a recently-invented aneroid barometer, which Glaisher wished to test. Coxwell had been instructed to take the balloon as high as he could and neither he nor Glaisher had any idea just how high it was possible for a gas-filled balloon to travel.

The first 20,000ft were fairly uneventful. It grew so cold that the men's lips turned blue, but other than that, nothing of note happened. It had been observed by mountaineers that the temperature dropped steadily with increasing altitude, so the fact that it grew very cold was no surprise. By 25,000ft though, their breathing was becoming laboured and they could hear the beating of each other's hearts. When they reached 30,000ft, Glaisher wrote in his record book that, 'it requires the exercise of a strong will to make and record observations'. They were by now five and a half miles high, on the edge of the stratosphere and wisely decided that they had had enough.

It was the flight which the two men made almost two months later which was to make them celebrities in Britain and across the world. At 1:00pm on 5 September, they took off once more from the gasworks. It took the balloon a little over three-quarters of an hour to rise 26,000ft and this was when things went terribly wrong. Glaisher was breathless and giddy, due to the thin atmosphere at that height and found that he was unable to read the thermometer. Turning to ask for help from Henry Coxwell, he was horrified to find that he was now apparently alone in the basket of the balloon. To his amazement, he saw that Coxwell had climbed out of the basket and was now struggling with the metal hoop which connected the balloon with the basket. The explanation for this alarming state of affairs was simple, although he did not learn until later what it was.

To release gas from the balloon and so descend, it was necessary to pull a cord, which operated a valve. In their rapid ascent, which had caused the balloon and basket to spin round and round, the cord had become tangled and could not be operated from the basket. Coxwell had therefore climbed up to try and free it. This was no easy task, and as he grasped the metal hoop and did his best to grab the cord and open the valve, it became so cold that his hands froze to the metal of the hoop. Being a resourceful man, he managed to grasp the cord in his teeth and by vigorously nodding his head, the valve ·was opened, gas released and the balloon stopped rising and began to head back to the ground.

Glaisher had fainted from lack of oxygen and when he came round, he found that Coxwell was laying on the floor of the basket, groaning. He told Glaisher, 'I have lost the use of my hands, give me some brandy to bathe them.' Of course, it was only to be expected that two Victorian adventurers should have taken a bottle of brandy with them, up into the stratosphere. Coxwell's hands were black from the effects of oxygen deprivation and frostbite, but luckily no permanent damage had been done. After rubbing brandy on his companion's hands and wrists, James Glaisher continued with his record-keeping.

Once they were back on the ground, the instruments were examined and Glaisher calculated that they had reached a height of just over seven miles above sea level; a record which was to stand until the middle of the twentieth century. Glaisher and Coxwell became famous for their exploits. There was one curious side effect of the journey and this was that the aneroid barometer was shown to be both accurate and robust. The usual type of barometer at that time measured air pressure by the level of a liquid. The aneroid barometer though had a metal box from which the air had been evacuated. This was connected by a spring to a pointer. As the air pressure rose, the box contracted and vice versa. It had only been invented a few years before the flights made by Glaisher and Coxwell and was regarded until then as a new-fangled device about which people were not really sure. Following the celebrated balloon flight, it proved its worth though and became popular in ordinary homes. The so-called 'banjo barometer', the archetypal feature of the hall of a well-to-do Victorian house, dates its rise in popularity from the flights of Glaisher and Coxwell.

It is interesting that Jules Verne's most famous novel mentioning balloons, *Five Weeks in a Balloon*, was published the year after the *Mammoth's* epoch-making fight. It is also interesting to note that while Verne's book was set in Africa, Henry Coxwell was actually commissioned to make a balloon which, it was planned, would be sent to Africa. The British Army were fighting the Ashantis in what is today Ghana, and it was thought that a balloon would make a grand addition to the expeditionary force. In the end, nothing came of the idea, because the practicalities of generating enough hydrogen for the balloon in the middle of equatorial Africa proved insuperable, but it is as neat an example as one could hope for, of fiction and real life becoming entangled.

Despite all the progress being made in ballooning, it was clear to some people by the 1880s that the future of aviation would be not with dirigibles and balloons, but rather with aeroplanes. A myth has grown up over the years that until the Wright brothers took to the air in 1903, heavier-than-air flight was widely believed to be impossible. This is not at all true. All the principles of flight had been established by 1850 and the only thing holding back the creation of viable aeroplanes was the lack of a suitable source of power. Once again, the French were leading the field.

Today, we use the term 'monoplane' as an ordinary noun, to distinguish aeroplanes with one set of wings from those with two, which we call 'biplanes or three, which are known as 'triplanes'. Few people realize that just as 'Hoover' is used as a generic term to refer to vacuum cleaners, rather than one specific brand, so too 'monoplane' is a particular name for one person's invention.

Felix du Temple was born in the Normandy district of France in 1823. He joined the French Naval Academy in 1838 and saw action in various theatres of war, including Mexico and the Crimea. We saw earlier that John Stringfellow's model aeroplane is generally regarded as the first to achieve powered flight, but the French have always believed that Felix du Temple has the stronger claim, even though it is agreed that his own model did not take off until almost 10 years later in 1857. They assert this, because unlike Stringfellow's plane, that of du Temple took off without any wires or ramps.

The first version of the miniature aeroplane used a clockwork motor, which was good enough to get the model off the ground, but obviously a full-size version would not be able to run on clockwork. For the next test, Felix du Temple and his brother Louis designed and made a very efficient little steam engine and this worked even better than the clockwork motor. As a precaution, Felix du Temple patented the design for a full-sized aeroplane, with dihedral wings and retractable landing gear.

It must not be supposed that Felix du Temple was working continuously on the problem of manned flight. His naval career occupied much of his time and it was not until 1874 that he was able to get around to building an aeroplane big enough to carry a pilot. He called this the 'Monoplane' and it was the first heavier-than-air machine to leave the ground under its own power. The Monoplane was extremely light, being made mostly of aluminium struts. Empty, it weighed just 176lbs and had a wingspan of about 40ft. To build up enough speed for the Monoplane to take off, it was launched from the top of a ramp, a little like a ski-jump. The steam engine was set going, the propeller span and the machine, with du Temple at the controls, roared down the ramp and then glided into the air for a short distance. Although sustained flight was not achieved, the machine definitely left the ground. The Monoplane was displayed at the 1878 World Fair in Paris.

It was, as the end of the nineteenth century approached, a matter of time before an aeroplane took off and flight was sustained. In Russia, 10 years after Felix du Temple's powered hop, Alexander Mozhaiski's steam plane took off and flew for a number of feet; the precise distance is a matter of dispute. This was in 1884. Six years later, the focus of powered flight moved back once more to France.

Clement Ader was an electrical engineer before he turned his attention to aeroplanes. From 1886 onwards, Ader carried out many experiments with airflow over wings, as well as trying to find the best and lightest steam engine possible. When he felt that he had perfected a new kind of aeroplane, which he named the *Eole*, Ader applied for a patent on 19 April 1890. Six months later, after some more fine tuning of the machine which he had been working on for so long, the moment came when it was put to the test. Unlike previous manned aeroplanes, the *Eole* was not reliant upon ramps or slopes. It had a wheeled undercarriage and took off in the same way as a modern aeroplane; that is to say by travelling along the ground, gathering speed until it had enough velocity to take to the air.

On 9 October 1890, Ader climbed into the *Eole* and started up the 20hp engine. The strange contraption had wings shaped more like a bat's than a bird's, but they served their purpose well enough. The aeroplane accelerated and moved faster and faster across the ground, before taking off and flying through the air for 165ft. This was considerably longer than the first flight made by the Wright brothers' *Flyer* on 17 December 1903, the occasion generally regarded as marking the beginning of true, heavier-than-air flight.

It must be said at once that there was no way of controlling the *Eole* in the air and that if the flight had continued for much longer or gone any higher than a few feet, then Clement Ader would have been in serious trouble; a crash would have been inevitable. Nevertheless, it is indisputable that it was Ader and not either of the Wright brothers who first left the ground in a self-powered aeroplane. A powered hop it may have been, rather than sustained and controlled flight, but there is not the least doubt that to Ader belongs the distinction of having taken to the air first, without the assistance of any ramp.

Hiram Maxim was an American inventor, born in 1840. There were few fields in which Maxim did not dabble. His inventions range from an inhaler to relieve asthma to a new sort of curling iron, from coffee substitutes to automatic sprinklers. In the late 1870s, Maxim installed the first electric lights in a New York building and then became involved in a legal battle with Thomas Edison over the invention of the incandescent light bulb. Perhaps Hiram Maxim is most famous for the invention of the machine gun which bears his name. The Maxim gun revolutionized warfare. According to Maxim himself, designing a machine gun had been undertaken after he had been told that this could make his fortune. He wrote;

> In 1882 I was in Vienna, where I met an American whom I had known in the States. He said: 'Hang your chemistry and electricity! If you want to make a pile of money, invent something that will enable these Europeans to cut each other's throats with greater facility.'

There had of course been machine guns before, the Gatling gun for example, but these had required the operator to turn a handle to feed ammunition into the breech. Maxim's weapon used the recoil of the firing to work a mechanism which cocked the gun again and allowed it to fire automatically. All that was necessary was to pull the trigger.

In 1881, Hiram Maxim came to England to organize the London offices of the United States Electric Lighting Company. He fell in love with the country and his visits home became more and more infrequent, until he decided to stay in England for good and acquired British nationality. Once he was settled on an estate near Bexley in Kent, Maxim decided to devote himself to the problem of powered heavier–than–air flying machines. If anybody could solve the difficulties which this entailed, the famous inventor was surely the man to do it. Maxim had long had an interest in powered, heavier–than–air flight. His own father had designed, although never built, a helicopter with twin rotors.

When the Wright brothers began their experiments with gliders and powered aeroplanes at the beginning of the twentieth century, they were careful to keep weight down to an absolute minimum, building their planes from spruce wood and fabric. This was not Hiram Maxim's way. From the late 1880s onwards, he constructed increasingly large aircraft, which he called test rigs. Describing them as 'large' understates the case dramatically; they were the size of modern airliners. The one which was finally tested successfully in 1894 had a wingspan of 125ft and weighed 3.5 tons. The Wrights' *Flyer*, by comparison, weighed less than a tenth of this. The propellers on Maxim's test rig were over 17ft long and the whole structure was built of steel tubing, with no consideration at all for keeping the thing light. It was powered by two enormous steam engines, each capable of providing 180hp. The total lifting area of the wings was around 4,000ft^2. The test rig may be seen in Illustration 13.

The whole aim of Maxim's project was not actually for his test rig to take to the air in free flight. He simply wished to prove that his theories were correct and that with sufficient power and a large enough area of cambered wing surface, that it was possible to lift such a large structure from the ground. The test rig ran along a broad-gauge railway track and arrangements had been made so that it could not rise more than a few inches, however successful the aerodynamic design should prove. Wooden guide rails ran alongside the railway track and these were intended to keep the test rig tethered to the ground when moving. Outriggers had small wheels which would engage the guide rails if the test rig should rise from the main rails. These rails would prevent the machine from flying up into the air.

After years of experimenting, both with the full-size version of the test rig and also models in a specially constructed wind tunnel, the day arrived when the great inventor was ready to show the public what his machine could do. Newspapers and magazines such as the *Times* and *Scientific American* were invited to attend the demonstration, as well as many important people, including the Prince of Wales, later to become King Edward VII, and also the author H.G. Wells. This was fortunate, because the presence of so many eyewitnesses ensures that there is not the slightest doubt about what happened on that summer day in 1894.

Tuesday, 31 July 1894 dawned bright and clear. After a couple of slow runs along the 1,800ft track, just to check that everything was running smoothly, the time came for the test of the machine's full capabilities. The engines were turned on to full power and the great leviathan shot off along the railway track at 40mph. On board were Hiram Maxim and two mechanics, Tom Jackson and Arthur Guthrie. As has been remarked before, weight was not a consideration in the way that it was with the Wright brothers' test flights.

Perhaps Maxim's own account of that memorable day is the best and most accurate. It was, remember, backed up by many witnesses, including cynical journalists who would have been only too happy to report a miserable flop and the failure of the famous inventor's enterprise. Maxim wrote,

> When everything was ready, with careful observers stationed on each side of the track, the order was given to let go. The enormous screw thrust started the machine so quickly that it almost threw the engineers off their feet, and the machine bounded over the track at a great rate.

Maxim went on to explain how he had increased the power from the steam engines and that the speed rapidly increased. He continued,

> When 900 ft had been covered, one of the rear axle-trees, which were of 2-inch steel tubing, doubled up and set the rear of the machine completely free. The pencils ran completely across the cylinders of the dynagraphs and caught on the underneath end. The rear end of the machine being set free, raised considerably above the track and swayed.

The other restrained wheels tore through the guide rails and the test rig rose into the air, travelling in this way at a height of between 6 and 8ft for a distance of several hundred feet. When Maxim and the mechanics realized that they were actually airborne, they at once closed down the power and the test rig sank back to the ground.

There are two curious points about the first and only flight of Hiram Maxim's colossal biplane. The first is that although both the *Times* and *Scientific American* reported it in detail and described it as the first successful flight of a heavier-than-air flying machine, it has been lost to history, supplanted in the minds of most people by the Wright brother's feeble efforts almost a decade later. We shall see later in this chapter why this might have been. The second point is the presence of H.G. Wells at the flight. We saw in Chapter 1 the association between Victorian writers of speculative fiction like Wells and Verne and their connection with modern steampunk. Both tie in neatly to real technological developments in the nineteenth century. Works such as *The War in the Air*, written in 1907 and containing vivid accounts of battles between aerial armadas were more likely to have been influenced by Wells witnessing the gigantic flying machine of Hiram Maxim taking to the air than by newspaper reports of the flimsy machines knocked up by the Wright brothers.

It is suspected that many readers will by now be asking themselves why they have been convinced since childhood that the Wright brothers were the first to build and fly a heavier-than-air flying machine when there were so many predecessors, notably Hiram Maxim's mighty biplane. To understand how history has been so grossly distorted, we need to look at another steam-powered aeroplane; the first heavier-than-air flying machine to achieved sustained, powered flight.

At about the same time that Hiram Maxim was experimenting with steam planes in England, the Secretary of the Smithsonian Institution, America's most prestigious group of museums, was doing much the same thing in the United States. Samuel Pierpoint Langley was a mathematician and astronomer and also a firm believer in the future of powered flight. Langley, who turned 60 in the year of Maxim's successful flight, designed an aeroplane and commissioned engineers to build a quarter-scale model of it, with a wingspan of 14ft. He christened this *Aerodrome* and several versions were built, each more efficient than the last. The twin screws were driven by a steam engine.

In 1896 two of Langley's aeroplanes were tested. They achieved stable and controlled flight over a distance of 4,200ft, an amazing achievement. Alexander Graham Bell was present at one flight and took a photograph of *Aerodrome* No. 6 soaring over the Potomac river. Two years later, Langley was given a government grant of $50,000 to build a full-size version of the *Aerodrome*, one capable of carrying a person. The tests of the larger version of Langley's plane, which took place in 1903, were all flops. Each ended in the aeroplane crashing straight into the Potomac on take-off. That same year, the Wright brothers carried out their first flights in the *Flyer*.

Before going any further, we might consider that the Wright brothers' claim to fame is quite tenuous and elusive. Others before them had flown in heavier-than-air flying machines, both powered and unpowered. Some had made powered flights of roughly comparable length to those that the Wright brothers made in December 1903. It is not always appreciated that the Wright brothers had chosen that spot and time of year for their flights because of the gale-force winds which howled along the beach there in the winter. In 1902 they had launched gliders which stayed aloft for longer than the *Flyer*, powered by nothing more than the fierce winds! The *Flyer*, which was essentially a glider fitted with a small petrol engine, might very well have been able to take to the air even without the assistance of the petrol-driven propeller.

The Wright brothers entered the history books for having undertaken the first manned, powered, sustained and *controlled* heavier-than-air flight. This is a little misleading though, because their flights in 1903 were all in a straight line, just as was Hiram Maxim's in 1894. He too had controls, although he, like the Wright brothers, did not really have a chance to use them, because the flight was so short. The Wright brothers were important pioneers of flight, but their claim to have built and flown the world's first aeroplane was, at least in the first few decades of the twentieth century, a little shaky.

The Smithsonian, while keen to display the *Flyer* in their museum, also championed Samuel Langley as the first man to build a successful aeroplane, on the grounds that his early, quarter-scale models flew and that his full-size version was actually capable of carrying a person. Langley was, after all, at one time Secretary of the Smithsonian and they clearly wished to give him a boost. They displayed Langley's aeroplane in the museum, with a sign declaring that it was, 'The first airplane capable of flight'. In 1918, Orville Wright – his brother Wilbur was by this time dead – took great exception to this and fell out with the Smithsonian. His irritation took the most practical form of removing the *Flyer* from the Smithsonian, where it was a valued exhibit, and having it shipped across the Atlantic to the Science Museum in London where it was to remain for 20 years.

The Smithsonian was desperately keen to have the *Flyer* back, so keen that in 1948 they were prepared to sign an extraordinary and legally binding contract which Orville Wright presented to them. Wright died in January 1948, before negotiations had been completed, but his executors duly presented to the Smithsonian the contract which he had drawn up. So keen were the museum to get the *Flyer* back, that they were prepared to agree to what was, in effect, a promise never to investigate the history of aviation. One passage reads as follows;

> Neither the Smithsonian Institution or its successors nor any other agency, bureau or facilities, administered for the United States of America by the Smithsonian Institution or its successors shall publish or permit to be displayed a statement or label in connection with or in respect of any aircraft model or design of earlier date than the Wright aeroplane of 1903, claiming in effect that such aircraft was capable of carrying a man under its own power in controlled flight.

For over 70 years, the Smithsonian, administered by the United States government, has been obliged to ignore any evidence of powered flight which predates that of the Wright brothers. This accounts, at least in part, for the fact that steam planes have rather been obscured from sight over the years and the *Flyer* promoted as the only contender for the honour. This secret agreement has had a chilling effect upon the Smithsonian Institution's readiness to examine objectively the history of manned flight. In recent years, for instance, more evidence has come to light relating to the possible achievements of a German mechanic called Gustave Whitehead, who was building gliders and steam planes in the Connecticut town of Bridgeport at the turn of the century. On his marriage certificate, dated 2 November 1897, Whitehead gave his occupation as 'aeronaut'.

According to witnesses, backed up by photographs, Gustave Whitehead flew a steam-powered aeroplane for a considerable distance in 1899 and in a letter to the *American Inventor* he claimed to have flown his latest aeroplane for a distance of two miles on 17 January 1902. The evidence is certainly sketchy and debateable, but the Smithsonian, who should be objectively interested in the veracity of such stories, dare not risk becoming involved in the controversy for fear of losing one of their prize exhibits.

There was a curious coda to the story of the steam plane and it came 30 years after the first flight of the Wright brothers. Readers have probably concluded that despite all the talk of Hiram Maxim's steam leviathan and the possible exploits of Gustave Whitehead, the future of aviation really lay with petrol engines. After all, a steam plane could never have been anywhere near as effective as one propelled by an internal combustion engine, could it?

In 1933, another pair of American brothers, George and William Besler, installed a very powerful steam engine in a biplane and showed that in many ways, their aeroplane was superior to those with engines fuelled by petrol. Not only did the 150hp engine enable the plane to perform aerobatics as well as any other biplane of the day, it was almost completely silent. This is because ordinary internal combustion engines operate by a series of explosions which, even when muffled, render them very noisy. The engine in the Beslers' plane was so quiet that when it swooped low above the crowd during a test flight in California, the pilot was able to have a shouted conversation with those below; something which would have been quite impossible in an ordinary aircraft. This silent engine might have made the Besler an ideal, early 'stealth' plane.

There was another feature of the Beslers' revolutionary aeroplane which set it apart from other planes of the time and that was that it could land in an astonishingly short space. This was because the steam engine allowed the propeller to change direction instantly and begin thrusting in the opposite direction. When the plane came in to land at 100mph, it was able to come to a halt in a distance of less than 100ft.

Our look at steam planes and metal airships has led us on a tour of many strange flying machines, none of which would have looked out of place in any steampunk story. In the next chapter we shall be examining a different aspect of the nineteenth century, mechanical computers and calculators. One of these, the difference engine invented by Charles Babbage, features of course in the title of the first modern steampunk novel.

Chapter 5

Steam-Powered Computers and Mechanical Calculators

W e often tend to assume that there is something relatively new about the handling and use of digital information. We talk blithely of 'digital' devices, often having no clear idea of what we even mean by the term. It is enough to know that there has been a digital revolution and that we now live in the digital age. There is, however, nothing startlingly modern about digital computers. The basic units of modern information technology, things like digital computers themselves, their central processors and memory, as well as the printers used to produce hard copies of what has been handled in the computer, were all devised almost 200 years ago. Not only that, but the technology used in those early digital devices was very robust, durable and efficient. Mechanical calculators were being used in offices well within living memory. In this chapter, we shall see that the digital technology of the nineteenth century was, until very recently, as fast and efficient as our own electronic devices; indeed, faster in some cases. The machines may have been made of brass, steel, glass and ivory, but they were as effective as many of their modern electronic counterparts.

One book is often cited as being the first, true steampunk novel; *The Difference Engine*, published in 1990. *The Difference Engine* is set in an alternative universe; Britain in 1855, in the full flow of an information revolution. The point of divergence from our own world is some time in the early 1820s, when a brilliant scientist perfects a huge mechanical computer. This initial invention, the difference engine of the title, is soon superseded by vastly more complex computers called analytical engines. These are powered by steam engines. The use of information technology at such a time causes upheavals in society, leading firstly to the imposition of martial law by Prime Minister Arthur Wellesley, Duke of Wellington, followed by a civil war and revolution. With the Industrial Revolution enhanced by an Information Revolution, British society is transformed into a meritocracy of intellectuals and scientists. This, in brief, is the plot of the novel.

There is something indescribably entertaining about the ideas explored in *The Difference Engine*. Perhaps the main attraction is the stupendously outlandish notion of computers which are driven by steam. It seems so obviously the stuff of fantasy. Perhaps it will therefore come as something of a surprise to discover that such mechanical computers, with provision for a man to tend the steam engine which ran them, were actually being planned in the 1820s and came very close to being built. Just as with other seemingly bizarre aspects of steampunk novels, the steam computer is rooted firmly in historical reality.

Charles Babbage was born in 1791 into a wealthy family. Even before he went to Cambridge University, young Charles had shown an inventive flair. He devised and constructed, for instance, a pair of shoes which enable him to walk on water; essentially a pair of wooden boards attached to shoes, like giant skis. These worked, for a short while, before Babbage toppled over and nearly drowned, because his shoes made swimming to safety all but impossible.

At Cambridge, Babbage made friends with John Herschel, who in later life became a famous astronomer, chemist and pioneer of photography. Charles Babbage's inclinations were more towards mathematics and he went on to become both a member of the Royal Society and also Lucasian Professor of Mathematics at Cambridge, a post later held by Stephen Hawking. Charles Babbage had various ideas, some revolutionary and others frankly bizarre. He felt persecuted by street musicians, for example, the people we today call buskers. He carried out the most careful calculations and concluded that he had spent a quarter of his life hearing the, to him, appalling noise of barrel organs and hurdy-gurdys. Because he was famous and influential, Babbage managed to get an Act banning street music passed by Parliament. This was counter-productive, because most people found the sound of street musicians a pleasant change from the hubbub of the Victorian city. Babbage was booed in the streets and people paid musicians to play all night outside his London house. One more example will perhaps indicate something of his character.

In 1842 the young Alfred Tennyson, later to become the Poet Laureate, published a poem called *The Vision of Sin*. The famous mathematician wrote to the poet, with an idea for improving the poem by making it more scientifically accurate. The target of his criticism was the couplet, 'Every moment dies a man, Every moment one is born'. We cannot do better than look at the letter which Charles Babbage wrote to Tennyson;

Sir:
In your otherwise beautiful poem 'The Vision of Sin' there is a verse which reads – 'Every moment dies a man, Every moment one is

born.' It must be manifest that if this were true, the population of the world would be at a standstill. In truth, the rate of birth is slightly in excess of that of death.

I would suggest that in the next edition of your poem you have it read – 'Every moment dies a man, Every moment $1^1/_{16}$ is born.'

The actual figure is so long I cannot get it onto a line, but I believe the figure $1^1/_{16}$ will be sufficiently accurate for poetry.

I am, Sir, yours, etc.,

Charles Babbage

Nobody knows if this letter was written with tongue in cheek or whether, on the other hand, poetic licence offended the scientist's feeling for truth.

Before seeing how and why Babbage became the father of modern computing, a slight diversion will be necessary. After Isaac Newton's death in the early eighteenth century, huge mathematical endeavours were undertaken in order to make sense of the universe by calculating the paths of planets and comets, and also predicting natural phenomena such as the rise and fall of the tides. As the Industrial Revolution gathered pace, others needed to use complicated mathematics to navigate ships, work out compound interest and a hundred and one other things. It is no exaggeration to say that science and commerce, from the seventeenth century onwards, relied upon increasingly complex mathematics. So too did the growing British Empire. Britain was a sea power and being able to deliver warships on time to the correct location was crucial in establishing British dominance at sea. This too, relied upon all sorts of mathematical work, backed up by meticulous observations of the sky. With no computers or electronic calculators, all this had to be done, by and large, by hand. There was one short-cut to lengthy calculations, but it introduced a new set of difficulties of its own.

Anybody who attended school before 1970 will know that carrying out intricate sums, multiplication for instance, usually required the use of books of tables; log tables, sine tables, tan tables and so on. For multiplying numbers, particularly very long numbers, one would look them up in a log table and then add the numbers together and find the anti-log. This would show you the correct answer to your calculation. If you wished to multiply 89 by 62, to give a very simple example, you looked up these numbers in a table of logarithms. There, you would find that the log of 89 is 1.949 and the log of 62 is 1.792. By adding these two figures, a total of 3.741 was obtained. Looking this up in another table told you that 3.741 was the log of 5518. This was the answer to the sum; 89 x 62. Essentially, the most difficult multiplication could be reduced to the addition of two numbers. Of course, it was vital that those tables should be reliable and accurate. The way in which they were compiled virtually guaranteed that they were not.

Extracting logarithms is a taxing and fiendishly difficult process. The most commonly used method is to represent any number as a power of ten; so the logarithm of 4 is 0.602. These are not rational numbers though, which means that one can never obtain a wholly accurate figure. The more decimal places to which a logarithm is calculated; the more accurate will be the results when it is later used. Some log tables were compiled to be accurate to 8 decimal places, others to 10 or 12. To carry out the reckoning necessary to obtain such accurate figures, teams of young men called 'computers' were employed. Our modern word for an electronic machine was being used centuries ago and referred until fairly recently to people rather than machines.

When a group of men had found what they thought was the correct answer to however many decimal places, then the result would be sent to the printers and included in a new table. The problem was that not only were there many errors of calculation in a team project of this sort, but the handwritten answers would also be misread by the men whose job it was to set up the type and turn them into books of tables. A 2 might be taken for a 7 and so on. As a consequence, all the tables of logarithms produced in this way were inaccurate to varying degrees. A mistake in one of these tables could have far more serious consequences than merely throwing out of kilter some attempt to work out the exact path of the planet Mercury's orbit. It could also end in a ship at sea missing landfall to take on water and provisions by many miles. As Sir John Herschel, a leading Victorian scientist and friend of Charles Babbage put it, 'An undetected error in a logarithmic table is like a sunken rock at sea yet undiscovered, upon which it is impossible to say what wrecks may have taken place.'

It was to Herschel, in the summer of 1821, that Charles Babbage made his famous comment, which first suggested the notion of a steam-powered computer. Babbage had been working his way through some tables and was dismayed to discover many errors; so many in fact that he thought that the mistakes rendered the log tables all but worthless. He exclaimed to his friend in exasperation, 'I wish to God these calculations had been executed by steam!' Steam power was the very epitome of efficiency at that time and it was hardly surprising that Babbage should have seen it as a remedy to human mistakes, much as in our own time we look to digital technology to solve some of our problems.

Charles Babbage's grand idea, which he fleshed out over the next couple of years, was for a gigantic, mechanical calculating machine. This would work out things such as logarithms and then print out the results automatically, thus removing all human error from the process of putting together log tables. Because of the possibility of vastly improved tables which would aid the Royal Navy in its navigation, the government was interested in the idea of infallible log tables and in June 1823, a meeting was arranged between Babbage and the then Chancellor of the Exchequer Frederick John Robinson.

Before going any further, it might be helpful to think a little about the problems of calculating by hand and why this was such an important matter for the British admiralty. These days, anybody with a mobile telephone is able to pinpoint his or her place on the Earth within a few metres. This is done by the phone exchanging messages with satellites orbiting the earth. Before GPS systems, finding out where you were at sea, where there are no landmarks, could be a complicated and time-consuming business. Take finding one's longitude.

Longitude means nothing more than how far east or west one is from London, through which, for historical reasons, the Greenwich meridian runs. All one really needs to know to calculate longitude is what the time is where you are and what time it is in Greenwich. Every hour further away from Greenwich, or the Prime Meridian as it is now called, equates to 15 degrees of longitude. Put like that, the thing seems absurdly simple, but it was anything but. Without accurate clocks, it was impossible to know what time it was in Greenwich and so this had to be worked out by means of what are known as 'lunar distances'. The apparent distance of the moon from various stars changes each day and if one could work out what these were at Greenwich and then compare them with readings from on board a ship, then it would be possible to find the difference between the two and so know how many hours and minutes the ship was behind or ahead of Greenwich time. Finding the time at sea could be done without a clock, by measuring the inclination of the sun above the horizon. From that, one could then find the longitude. The only problem was that these calculations and measurements took about four hours.

In the middle of the eighteenth century the then Astronomer Royal, Nevil Maskelyne, began preparing and publishing a Nautical Almanac, which contained, for months ahead, the lunar distances at Greenwich for any given day. This reduced greatly the mathematical work which needed to be carried out on board the ship. Working out the lunar distances months in advance was an incredibly tedious business and the Board of Longitude, which was responsible ultimately for the almanac, employed schoolteachers and clergymen do the donkey-work. These individuals were known as 'computers'. Even when accurate and reliable chronometers began to appear, they were very expensive. Typically, a chronometer would cost about a third of the price of an entire ship. For that reason the cheap and readily available Nautical Almanacs remained in use well into the nineteenth century. It was not until the 1850s that they fell into disuse.

It was for complex and dull mathematical work of this sort that Charles Babbage's difference engine might have been invaluable. Needless to say, that army of schoolteachers and old clergymen, dedicated as they might have been, made many mistakes in their working out, something which would have been impossible for Babbage's computer.

Unfortunately for both Charles Babbage and the British government, no minutes were kept of the meeting between the eccentric scientist and the man in charge of the country's finances. After it was over, the two men had very different ideas about what had been agreed. What was never disputed was that the Chancellor of the Exchequer had agreed to hand over £1,500 in connection with Babbage's marvellous mechanical computer. As far as the Chancellor was concerned, this was the total cost and for that sum, the government would receive a completed model of the new machine. Babbage, on the other hand, thought that he had made it clear that the £1,500 was merely a down-payment; a preliminary amount which would enable him to begin work on the project.

£1,500 was a substantial sum of money in 1823, at a time when an agricultural labourer might earn perhaps £20 a year. One can sympathize with the Chancellor of the Exchequer for thinking that such an amount might buy him a new kind of machine for aiding in the navigation of the navy's ships. On his side, one can see why Charles Babbage believed that £1,500 would merely lay the groundwork for what he hoped to achieve. There had been mechanical devices before which could add numbers, but nothing as ambitious as this.

Pocket calculators date not from the 1970s, as many people suppose, but were being produced across Europe in the seventeenth century. One of the earliest models was that made in 1645 by Blaise Pascal, the French mathematician. The German philosopher Leibnitz invented another kind of calculator 30 years later. In England too, pocket calculators were being produced. The Science Museum at South Kensington contains an example made in London in 1666. This beautiful little machine is about the size and shape of a present-day smartphone and is made as precisely as a clock. The brass case and tiny dials have a decidedly steampunk air about them. Samuel Morland's calculator was designed to deal with the British monetary system at that time, which is to say 12 pennies to the shilling and 20 shillings to the pound.

By the time that Babbage launched his project of building a difference engine, mechanical calculators like those of Morland, Leibnitz and Pascal were well-known and the mechanisms for things such as carrying numbers and converting 10 units to a single lot of tens were familiar to many people. Babbage was building upon tried and tested technology.

There is all the difference in the world between a mechanical calculator the size of a smartphone which can add and subtract, and the fiendishly complicated and enormous machine which Babbage hoped to build. The engineering involved in the project required cogwheels to be made to very precise sizes and it was essential that each should be perfect or the working of the entire machine would be compromised. New machine tools had to be built simply to make the components of the difference engine. The costs spiralled and, having invested already a large sum of money, the government was at first reluctant to abandon what appeared to be a very useful piece of work; a machine which would make all sorts of calculations without any possibility of error creeping in. Not only that, it would actually print out these calculations as well.

Having sunk thousands of pounds in the construction of the difference engine, successive governments were reluctant to abandon the project and, little by little, handed over ever larger amounts from the Treasury. In 1829 the then Prime Minister, the Duke of Wellington, went to see a model of what Babbage hoped the difference engine would look like if completed. Wellington was impressed and arranged for a further £3,000 to be given towards the project. Five years later, with the difference engine still not completed, Babbage told Lord Melbourne, who was now Prime Minister, that he wished for more money towards an entirely new machine which he thought might be possible. This was the vastly more complicated analytical engine, which would be 8ft tall and weigh as much as a railway locomotive! Not surprisingly, he was told that until his difference engine had been completed, there would be no money available for any new ideas. After 11 years, it was looking to many people as though Charles Babbage was never going to produce anything practical, no matter how much money he was given from the public purse.

The curious thing about the proposed analytical engine was that being so large, it would have been impossible to turn the various cogs and wheels by hand. Babbage drew up plans which showed that this would be a steam-powered computer, just like those in the novel *The Difference Engine*. Rather than just a crank which could be operated by a single person, the enormous analytical engine would have had to be connected to its own steam engine and thus the resultant calculations would, just as Babbage had suggested wistfully in his letter to John Herschel in 1821, be 'executed by steam'. Alas, in the real world, it was not to be. The money for the project was about to be cut off. Computing by steam was to remain a dream.

For the better part of 20 years, Charles Babbage worked on his difference engine, building small parts of it and, at the same time, dreaming of the even more elaborate and ambitious analytical engine. This would be able to do far more than just work out logarithms. As he planned it, the analytical engine would have all the features of a modern computer, although the names used for some of the parts were different to those with which we are today familiar. In addition to the printer, there would be the 'mill', which undertook the working out and was roughly the equivalent of the microprocessor in a modern laptop. There would also be a 'store', where the results of calculations would be stored. This corresponded to the memory on an electronic computer.

In 1842 Babbage, after having spent thousands of pounds provided by the treasury, in addition to a huge amount of his own money, had still not produced a completed machine. One can hardly blame the government for deciding at this point that enough was enough and they pulled the plug on the difference engine by refusing to pay any more towards its development. The British government had by now given Charles Babbage an incredible £17,000 and, in effect, received nothing in return but promises of future wonders. Small sections of the difference engine had been built, but nothing that was of any practical use. Babbage was a notoriously difficult man to work with and had a habit of falling out with people, even his supporters. Along the way, he had made quite a few enemies. One of these was a secretary of the Royal Astronomical Society, the Reverend Richard Sheepshank.

Richard Sheepshank thought that Babbage had effectively been living off the taxpayer for nearly 20 years, while moaning the whole time about not being given enough money. So strongly did he feel about this, that Sheepshank wrote and published a book attacking Charles Babbage. It was called *Letter to the Board of Visitors of the Greenwich Royal Observatory, in Reply to the Calumnies of Mr. Babbage* and contained a brilliantly succinct summary of what many people by then felt about Babbage and his famous difference engine. He wrote, 'We got nothing for our £17,000 but Mr. Babbage's grumblings. We should at least have had a clever toy for our money.'

In *The Difference Engine* of course, history turns out rather differently and the difference engine of the title, really the analytical engine which Babbage envisaged, was an enormous success which revolutionized Victorian society and kickstarted the information revolution well over a century before it took place in our own world. One of the important characters in *The Difference*

Engine is somebody who worked closely with Babbage on his mechanical computers in the real world. Ada Lovelace was the daughter of the Romantic poet Lord Byron and she had a great interest in mathematics and science.

Although interest in Britain in Charles Babbage's work was sometimes lukewarm, there were those in Europe who could see how tremendously significant were the ideas which Babbage was trying to put to work in a practical way. The Italian mathematician Luigi Menabrea wrote an article on the difference engine and Ada Lovelace, who by that time had been visiting Babbage and discussing his work with him at length, translated this into English. To her translation, she attached some notes of her own, which were three times the length of Menabrea's original article. In these, she tried to give some idea of the limitless possibilities of the analytical engine, should it ever be built. Among these notes was a method for using the analytical engine to calculate Bernoulli numbers (a sequence of rational numbers which occur frequently in number theory). The sequence of instructions has been described as the world's first computer programme.

One of the most radical ideas which the analytical engine featured was that it would be programmable by means of punched cards. There was, of course, nothing startlingly original about the use of punched cards or tapes to give instructions to a machine. As early as 1804 a Frenchman called Jacquard had invented a loom controlled by a sequence of punched cards, which would contain the finished pattern.

For most of the twentieth century, computers were programmed by means of punched paper cards and tapes of the kind which Ada Lovelace had envisaged being used to programme Babbage's proposed analytical engine. This method of storing information and then allowing it to be read by machines was widely replaced from the 1980s onwards by magnetic discs and then flash drives. It lingered on though and was still being used in America as late as 2014 in the Votomatic machine used to record votes in an election. Attention was of course drawn to this antiquated means of processing data during the 2000 presidential election, which resulted in some of the IBM data cards not being punched correctly.

The widespread use of punched cards and tapes for information storage and computer programming was not adopted in the United States as a result of Ada Lovelace's writings on the subject. In America, punched cards of this kind originated in the Wild West and were first used to tackle a very nineteenth-century American problem; that of banditry on railway trains.

It is an image familiar to us all; train robbers in the Old West. With handkerchiefs pulled over the lower halves of their faces, they ride alongside the thundering locomotive, firing at the driver and forcing him to bring the train to a halt, so that they may rob the passengers. This was certainly one method of going about the thing, but there was another, simpler way. This entailed the bandits simply buying tickets like any other passenger and then taking over the train at gunpoint, before bringing it to a halt at a spot where their confederates would be waiting.

In an attempt to combat this kind of crime, an unknown person came up with the following scheme. What if a brief description of every passenger buying a ticket were to be both appended to their ticket and a duplicate record stored at the ticket office? Then, if the train were to be robbed and any passengers were to be in league with the criminals and leave the train with them after the robbery, then the railway company would have a description of the 'inside men' who had travelled on the train. This at least was the theory. The 27 August 1887 issue of the American magazine *Railway News* carried a description of how such 'punch photographs' as they were called worked. Each ticket and corresponding record of sale had the following printed at the side. The words were perforated, so that those which were inapplicable could be punched out;

Male - Female.
Slim - Medium - Stout.
Young - Middle-aged - Elderly.
Eye. Light - Dark.
Hair. Light - Dark.
Beard. Moustache - Chin - Side - None.

So it was that after a robbery, a suspect might be known to be a slim, young man with dark hair and a beard.

The 'punch photograph' was never widely adopted, but it suggested to a young German-American engineer a very interesting idea. Collating large amounts of data by hand was a laborious process and because it was so monotonous, those carrying it out were likely to lose interest and make mistakes. This was seen during the 1880 American census, the information from which was collected and listed entirely by bored clerical workers. Herman Hollerith had the notion of automating the procedure, so that no human error could creep in. The numbers would be counted by machine and so too would various things such as the differing numbers of men and women, the ages, number of members of each household and so on.

Once all the information had been placed on punched cards, it would be analysed by an electric reader which Hollerith had invented. This may be seen

1. This 1829 drawing shows a proposed 'Hyperloop'; a vacuum-powered transport system between Britain and India.

2. The 'pneumatic' railway operating in South London in 1867. Some pneumatic trains in Victorian England were faster than their modern-day electric equivalents.

3. Contemporary artist's impression of the world's first airliner, which was due to go into operation in the 1840s.

4. The world's first mobile phone; an 1870s 'radiophone'.

5. Worries about the depletion of fossil fuels caused some inventors in the nineteenth century to turn to renewable energy. This is a solar-powered printing press.

6. The Steam Man of the Prairies; a steampunk fantasy from 1868.

7. A steam-powered bus operating in London in 1838, the year that Victoria ascended the throne.

Left: 8. An optical telegraph; the origin of the code used for text on the Internet.

Below: 9. Not a death ray from a steampunk graphic novel, but a solar power plant at the 1878 Paris Exposition.

10. Skyping with a family member on the other side of the world in 1878.

Above left: 11. Perhaps the strangest mode of travel ever seen in Britain; the ocean-going electric tramway between Brighton and Rottingdean.

Above right: 12. The 1890 American census; the first ever to be tabulated electrically.

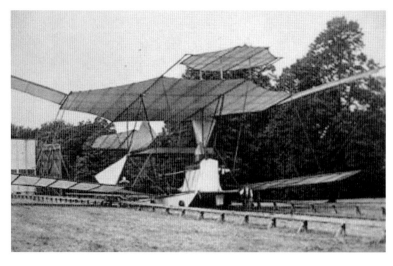

13. Hiram Maxim's steam-powered aeroplane, which took off in Kent in 1896.

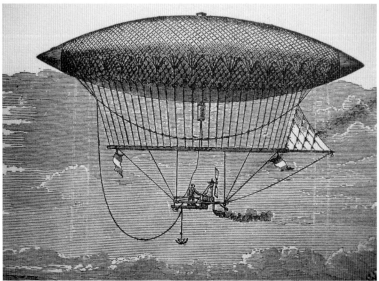

14. The first steam-powered airship in 1852.

15. The 11-mile-long shopping mall which was never built; it would have girdled central London on both sides of the Thames.

Above left: 16. A Victorian fax machine; surprisingly, fax machines pre-dated the invention of the telephone.

Above right: 17. The former headquarters of the optical telegraph chain which sent messages across Britain at hypersonic speeds.

Right: 18. American astronauts splash down after the first voyage around the moon . . . in 1865.

Left: 19. A steampunk dream; the interior of a spaceship, resembling nothing so much as a Victorian gentleman's study.

Below: 20. A Victorian desktop computer: the Scheutz difference engine.

in Illustration 12. If ever a piece of Victorian machinery looked as though it would fit neatly into a steampunk graphic novel, this is it. The tabulator was simplicity itself to operate. Each card, with various holes punched in it, was placed on a reader and a number of 'feelers', which were no more than electrical wires, were then lowered. Obviously, these could only go through the parts of the card where there was a hole. They then descended into little wells of mercury and completed a circuit. This had the effect of advancing one of the 40 dials on the display panel by one place. As the succession of cards were placed in the reader, the information was automatically added to the previous quantity and shown on the appropriate dial. At the end of each working day, the totals on the various dials were recorded and they were set back to zero for the next day.

Herman Hollerith's machine proved its worth when used to handle the results of the 1890 census. The information gathered in the course of the 1880 census had taken 18 weeks to collate. Over the next 10 years the population of the United States increased from 50 million to 60 million and yet it took Hollerith's tabulators just six weeks to accomplish what had in 1880 taken three times as long. There was little scope for human error, either. As *The Electrical Engineer* magazine remarked at the time, 'This apparatus works as unerringly as the mills of the gods, but beats them hollow as to speed.'

The difference engine which the British government paid Charles Babbage to build may have come to nothing in the end, but other difference engines were not only constructed at that time, but were found to be very useful. Inspired by what he had read of Babbage's work, a Swedish inventor called George Scheutz, working with his son Edvard, built a working difference engine. Like Charles Babbage, they applied for government funding for their work, but the Swedish government was a little more canny with its money than the British and declined to finance such a speculative scheme. By 1843, a few months after the British Treasury had finally called a halt to expenditure on Babbage's difference engine, the Scheutz father and son team demonstrated their machine to the Royal Swedish Academy of Sciences, who were vastly impressed with what they saw.

The Swedish difference engine could do everything which Babbage had claimed for his own invention; that is to say carry out mathematical calculations and then print the results. It was powered, like a pendulum clock, by falling weights and was about the size of an upright piano. With the seal of approval of the Royal Academy of Sciences, George Scheutz applied again to the Swedish government for money to help him improve the invention. This time, they agreed, but showed themselves a great deal more sensible than the British government had been with Charles Babbage. George and his son were granted a modest amount of money, but it was given on condition that a completed model of the machine would have to be provided within two years, failing which the money would have to be repaid.

The Scheutz difference engines, or tabulators as they were also known, proved to be a commercial success. One was sold to the Dudley Observatory in New York, where it was used to compute the orbit of Mars. The British government also bought one, which must have been very galling for Charles Babbage. This was used by the General Register Office to produce and print life tables. A Scheutz difference engine may be seen in Illustration 20.

Many readers will assume that all this talk of mechanical computers is by way of being a quaint visit to a remote and forgotten past. It is nothing of the sort. The creation of the difference engines in the middle of the nineteenth century spurred on research and development of other mechanical calculating machines which were so effective that they were still being widely used in offices in Europe and America, well within living memory. Before looking at this subject, let us look at other computers which were also being produced at around the same time as the difference engines.

One of the most important pieces of information to know if, like Britain in the nineteenth century, you were a seafaring nation and had an empire which stretched from one end of the world to the other, was the precise time at which the tide would ebb and flow. This knowledge is vital, whether you wish to run a port efficiently or launch a seaborne invasion fleet. It can be of critical importance, as was seen when the Spanish Armada appeared off the coast of England in 1588 or during the planning of the D-Day invasion of Europe in 1944. We all know the story of Francis Drake playing bowls and being unwilling to end his game to go rushing off to fight the Spaniards immediately. A classic case of English *sang froide*? The real reason that there was no particular hurry was that the tide was coming in when the Spanish fleet was sighted and so it would be some hours before it would be going out again and the English ships could leave port.

The problem with the tide is that it does not follow any easily recognizable pattern, but changes hour by hour and day to day according to a dozen different rhythms. Working these out by hand, the only method available until 1872, was a very complex and time-consuming process. That year. Sir William Thompson, later to be Lord Kelvin, found a way of arranging a dozen pulleys which would convert rotary motion into sinusoidal curves. Once a pen was added and a sheet of paper, this Heath-Robinson gadget was able to predict the tide for months in advance.

In a sense, *The Difference Engine* was right about Babbage's work providing the impetus for an information revolution in Victorian Britain. Charles Babbage's difference engine may not actually have been built, but the influence of the idea of a machine which was able to think, or at least carry out mathematical operations, was pervasive and led not only to the invention of tide-predicting machines but the adoption of calculating machines in offices.

The difference engines which were manufactured by George and Edvard Scheutz were huge machines which took up a large part of a room. They were clearly not intended for everyday use. There was, however, a market for simpler devices; mechanical calculators which could be operated by clerks in an office, say. There had been handy calculating machines before, but these had been fiddly and each one had to be made by hand. This meant that they were expensive novelties. In 1851, a desktop machine began to be mass-produced. The Arithmometer was invented by Frenchman Thomas de Colmar in 1820, but it was not until he set up a production line for the things, giving each a serial number, that they began to sell in large numbers.

The Arithmometer was a robust and easily-used calculator which could carry out all four of the basic, arithmetical operations; that is to say addition, subtraction, multiplication and division. It was the beginning of a revolution, because the Arithmometer was soon being bought by banks and insurance offices, who found that the machine saved countless hours of painstaking calculations done by hand. For the next few decades, the Arithmometer was the standard by which all other calculators were judged.

Inevitably, enterprising businessmen dismantled Thomas de Colmar's Arithmometer and began producing their own versions. At the same time, the quest was on for a way of making the use of these machines even simpler. After pressing the keys, calculating machines invariably needed to have handles turned or levers pulled to operate the mechanism. This of course slowed down operations a little. What was needed was a machine which was 'key driven'; that is to say that the simple action of pressing the keys would be sufficient in itself to work the mechanism. It was to be 1886 before an American, Dorr E. Felt, patented the Comptometer, which was still in use over a century later.

The Comptometer, which over the decades became a generic noun, rather like 'Hoover', had eight columns of keys. Each column consisted of nine keys and at first glance, the device looks very complicated. Once an operator had the hang of it though, his or her fingers would fly over the seventy-two keys at lightning speed, just like somebody touch typing. The Comptometer really was a product of the information revolution triggered by Babbage and his failed difference engine. Every office or shop could afford to have a Comptometer and they were, by the end of the nineteenth century, almost as ubiquitous as electronic calculators were to be a century later.

Mention of electronic calculators brings us to a very curious point about this piece of Victorian engineering and that is that, incredible as it may seem, the purely mechanical Comptometer was faster for a trained operator to use than a modern calculator. The reason for this is that when using an electronic calculator, it is necessary to press one key at a time. Accidently pressing two keys together will lead to an error. Not so with the Comptometer. Because of the way that the keyboard is arranged on the Comptometer, it is perfectly possible to enter a four-digit number by pressing four keys in different columns simultaneously. This, together with other short cuts which a skilled operator used without thinking, meant that for basic arithmetical operations, the electronic calculator is slower than using the mechanical calculator. It is not hard to see why.

Anybody who has watched somebody touch-typing and then compared it to their own feeble and painfully slow efforts to enter text on a standard keyboard will know that one is incomparably faster than the other. The method which many of us use, whereby we hunt for the letter we want and then jab a finger down at it, is known as 'Hunt and Peck'. When electronic pocket calculators first began to appear in offices in the 1970s, it was initially supposed that they would spell the end of the Comptometer and its skilled operator. What actually happened was that people soon found that they used 'Hunt and Peck' to input numbers on their calculators and that they often pressed two keys together or pushed the wrong button or managed to erase the answer inadvertently. Using an electronic calculator was far slower for simple arithmetic than having a proficient person use a Comptometer. Although Comptometers were no longer manufactured from the 1970s, some older people continued to use them and even as late as the 1990s they were to be found in American offices. It was the advent of the desktop computer which sounded the death knell for mechanical calculators, rather than the coming of the pocket calculator.

Chapter 6

Radiophones, Fax Machines and Hard Drives

When we think about telecommunication in the nineteenth century, the image most likely to spring to mind is probably primitive telephones of the candlestick type, with an earpiece separate from the main body of the phone. What else could there have been? It will come as something of a surprise to learn that much of the information and communication technology which underpins the world of the twenty-first century was invented while Victoria was on the throne and that the Internet itself depends heavily upon a system first used in the 1880s. Surely only a steampunk story would show Victorians using wireless telephones which transmitted speech via modulated beams of light? The same thing must go for fax machines too. Any novel showing a fax machine being operated in 1861 between Moscow and St Petersburg must be a fantasy; after all, the telephone wasn't even patented until 15 years later!

In fact, neither of the above ideas are fiction. Let us begin with wireless telephones. In Illustration 4 we see a very early model of a telephone which does not need to be connected with wires. Alexander Graham Bell, inventor of course of the telephone, called this device the 'photophone' and thought that it was of far greater importance than the familiar sort of telephone which he patented in America in 1876. This strange-looking device, which has a distinctly steampunk look about it, was actually the first step in the technology which underpins today's Internet.

The Internet may be accessed of course by radio. Much of the data though is carried on fibre-optic cables which run under the oceans. Without these cables, which carry a tremendous amount of information, the Internet and worldwide web would struggle to cope with the demand upon it. Pictures, spoken words, text and moving images are broken down into digital form and transmitted along the cables as pulses of light from lasers. These are then unscrambled and turned back into comprehensible information at the other end. A lot of telephone calls and television programmes are also sent along cables in this way, the principle being precisely the same. The information is broken down into modulated beams of light which then travel along a very thin fibreglass cable, to be decoded at their destination. Traditional telephone lines do much the same thing, only it is electric current which is varied instead of light, causing a miniature loudspeaker in the earpiece to reproduce the voice of the person speaking into the microphone at the other end. It was this system which Bell invented and for which he was granted a patent.

In 1878, when the telephone was still an exciting novelty, a man in London called A.C. Ball carried out a curious experiment. He arranged a microphone so that instead of sending electrical signals, it would cause a very thin mirror to vibrate. He found that if a beam of sunlight was now reflected from this mirror onto a plate coated with lampblack or carbon, it was possible to use a telephone earpiece and actually hear the words sent along the ray of light. Ball regarded this discovery as little more than an amusing party trick, seeing no practical or commercial applications for it. When Alexander Graham Bell heard about it though, he became very excited and set out to produce an improved and commercially-viable version of his own. At first, he too used a coating of carbon as the receiver and found that the speech received was, to use his own words, 'painfully loud'. Bell began to explore the possibilities of replacing the carbon with selenium.

Selenium is a non-metallic, semi-conducting element which was first isolated in 1817. It exhibits a very strange property. In darkness, selenium will not conduct electricity at all. When light shines on it though, it will conduct electricity in direct proportion to the amount of light present. This behaviour of selenium proved useful not only in communicating by light beams, but also in early mechanical television systems.

Having made a selenium receiver, rather than just a metal plate coated with soot, Bell found that it was possible send messages over a considerable distance. By having a selenium cell at the focal point of a parabolic reflector, it was found that no distortion of the signal occurred at all. He was most excited by these developments, more so than he had been when he finally managed to get the first telephone to work. After his experiments with what he christened the 'photophone', Bell wrote to his father, saying, 'I have heard articulate speech by sunlight! I have heard a ray of sun laugh and cough and sing!' So enchanted was he with this new invention, that Alexander Bell wanted to name his new daughter 'Photophone'. His wife though would have none of it and stood firm for a more traditional name. The child was eventually called Marian.

There were two chief reasons for Bell's great enthusiasm for this new means of sending the human voice a great distance. The first of these was that the quality of reproduction in early telephones, those using copper wires, was atrocious. There was crackling, buzzing, humming, all manner of distortions and often the line went entirely silent. This was very irritating and the completely clear transmission of speech over the reflected rays of light made a startling contrast to what was possible even with the best line carried along wires.

The other great advantage of the Photophone, if it could be developed as a commercial proposition, was that it would enable the tangle of wires which were now starting to be seen above American cities, to be entirely done away with. Hundreds of wires were dangling from buildings and these all needed to be maintained. As well as being unattractive to the eye, their upkeep was labour-intensive and every time somebody wish to be connected to the telephone network, it meant a man clambering up a pole with a coil of wire and linking the new subscriber to the system with a new cable running to their eaves. Imagine how simple things would be if the only connection necessary were to be a parabolic dish fixed on the side of a house, rather like the aerials for satellite television? The idea of Victorian houses having such reflectors attached to the front of them is an enchanting one.

The obvious drawback to the system, which in retrospect will be seen as an insurmountable obstacle, was the weather. Using the photophone required not only bright sunshine, but also a clear and uninterrupted line of sight from transmitter to receiver. Even something as trifling as a shower of rain or smoke from a bonfire could be enough to disrupt communications. It was this which ultimately doomed the photophone. Experiments with mobile systems, as seen in Illustration 4, continued throughout the nineteenth century, but the requirement for bright sunshine really made the whole thing hopelessly impractical. It is seldom enough in America or Europe that one may be assured of constant and uninterrupted sun. This was to say nothing of the difficulties which would ensue at night.

Experiments were conducted in other countries with this mode of communication, but all were, in the end, failures. 'Photophone' was Bell's name for this device, but elsewhere it was known as the radiophone; a startling foreshadowing of the future in a world where radio transmissions were not yet known. Later in the century, artificial sources of light were used. Carbon arc lamps seemed promising at first, being very powerful, but the light faded and dissipated after a short distance. It was not until the invention of the laser that it became a practical proposition to use light for sending telephone calls across great distances.

Sending telephone calls along rays of light is an astonishing thing to find happening in 1880. Perhaps even more surprising is the discovery that fax machines were invented not in the twentieth century but as early as 1843, just five years after Victoria came to the throne. This was not only more than 30 years before the telephone was patented, but it was at a time when even electric telegraphs were an exciting, cutting-edge means of communication

Alexander Bain was the poorly-educated son of a Scottish crofter. He was possessed, though, of a brilliant mind and daring imagination. As a youth, he constructed a delicate clock from bits and pieces which he found lying around. The spring powering it was made of a coiled twig of heather. After a very sketchy and rudimentary education, Bain was apprenticed to a clockmaker. By the time he was 20, he was drawing up plans for all manner of futuristic devices. Alexander Bain invented something which we take so much for granted today, that it is all but impossible to imagine a world without it; accurate and nationwide timekeeping.

After inventing the electric clock, one whose pendulum was driven by an electromagnet rather than the more conventional coiled spring or hanging weight, Bain came up with the idea of a 'master and slave' system of electric clocks linked by telegraph wires. As the age of the railway gathered pace in Britain, the fact that different towns and cities were keeping different times became a serious problem. Bristol, for instance was 10 minutes behind London. For stagecoach journeys, such a difference in time is trifling. For two trains sharing a track at 10 or 15-minute intervals, it can be quite literally disastrous. Alexander Bain devised a method whereby the pendulum of one electric clock in London would, as it reached one end of its swing, send an electric signal to a clock in Manchester, say. This would mean that both clocks were synchronized and in perfect step.

Having found a way of ensuring that two pendulums hundreds of miles apart were in harmony, with both their pendulums swinging in unison, Alexander Bain had another idea. How if the swinging pendulums of his synchronized electric clocks could be used to send information from one place to another? In 1843, he applied for a patent relating to, 'improvements in timepieces, and in electric printing, and signal telegraphs . . .'. He went on to describe how, 'a copy of any other surface composed of conducting and non-conducting materials can be taken by these means'. Hidden within these vague words was the idea for a mechanical fax machine.

Although extremely crude, Bain's idea was that any printing or images could be turned into plates, parts of which would conduct electricity and other parts of which would not. If an electric stylus were to swing back and forth across such a plate, then it would send a message to a distant counterpart when the material it passed across was able to conduct electricity and then stop sending a current when it came to a part of the plate which did not conduct electricity. If a similar stylus at the other end of the link were to be swinging over paper treated with chemicals which would darken when an electric current was passed through them, then text or images could be reproduced there, having travelled along ordinary telegraph wires. In short, the fax machine had been invented, 33 years before Alexander Graham Bell would apply for a patent for his telephone.

In America, Samuel Morse was hoping to keep any developments relating to telegraphs in his own hands and a legal battle resulted, leading to the American courts ruling in Morse's favour; that Alexander Bain's idea was an infringement of patents already held by Morse. Others in Britain though were working along the same lines as Bain and at the Great Exhibition, held in London in 1851, a man called Frederick Bakewell demonstrated a much more efficient method for transmitting images. Instead of a swinging stylus, he used synchronized rotating cylinders. As with Bain's version, chemically-treated paper, which would change colour when a current was passed through it, was used in the receiver.

Although Bakewell's machine was capable of sending clear images through wires, the process was agonizingly slow and also suffered from difficulties in precisely synchronizing the sending and receiving apparatus. It was left to another inventor to refine these methods and turn them into a commercially viable system. An Italian physicist called Giovanni Caselli both speeded up the process of sending images and also made it far simpler. Pictures or text were printed in non-conducting ink on a thin sheet of tin plate. This was scanned by an electric stylus. Caselli christened his new and improved fax machine the 'pantelegraph'. In 1860, Caselli was invited by Napoleon III, the emperor of France, to show off his invention. He did so by arranging for the signature of the celebrated composer Rossini to be sent from Amiens to Paris, a distance of 86 miles. Hugely impressed, the emperor asked if the device could operate at greater distances, and so Caselli had a message sent to Lyons, which is almost 250 miles from Paris.

The French were, in the eighteenth century, as we saw in Chapter 2, pioneers of the optical telegraph. Now, under the encouragement of Napoleon III, they were equally enthusiastic about the fax machine. At the emperor's urging, legislation was passed for the setting up of a public fax line between Paris and Lyons. In its first year, almost 5,000 faxes were sent along the line, the main use of the service being for verifying signatures for banking purposes. Any member of the public though could use the line, at a cost of 20 centimes per square centimetre of an image. The service was extended to Marseilles and was used regularly during the 1860s. Tsar Nicholas I heard about the French fax service and for a year, between 1864 and 1865, had an experimental line installed between his palaces in Moscow and St Petersburg.

The Franco-Prussian War of 1870 brought about the destruction of many telegraph lines in France and after the war ended, the fax service was not reinstated. Developments after this time took place chiefly in the United States. Elisha Gray, unlike Alexander Graham Bell, is not a familiar name to modern readers, which is strange, because he invented the telephone at almost exactly the same time as his better-known rival in the field. Both Gray and Bell filed their patents on the same day; 14 February 1876. Only a few hours separated the two applications and so it is the first one which we recall today and the man who came second has been all but forgotten. Elisha Gray continued to invent electrical equipment for use in communications though and in 1888 he was granted a patent for the 'telautograph', which sent images through telephone and telegraph lines by scanning both horizontally and vertically. This greatly improved the quality of the image being transmitted.

In fact, fax machines were being widely used in late nineteenth-century America. Illustration 16 shows one from the 1890s. At the bottom right-hand corner is an example of an image sent through this particular system.

We have seen that fax machines were in use long before the invention of the telephone, which might come as a surprise to many readers. Once again, pictures of such things, as in the one shown in Illustration 16, would not look at all out of place in a steampunk context. We come now though to something which will, on the face of it, appear to be wholly unbelievable; the granting of a patent for television in 1885. This clashes with all that we think we know. Didn't John Logie Baird invent television after the First World War? The roots of television lay a lot further back in the past than this.

The appeal of steampunk lies in the way that alternative universes are created, in which either the Victorian world or that of the modern day are altered so that the one is more advanced than we expect or the second retarded and more backwards than the world with which we are familiar in real life. So in a novel like *The Difference Engine*, we encounter a version of 1850s England which has racing cars and computers. Conversely, in an Alternative Universe proto-steampunk novel such as *Pavane*, the world of the early twenty-first century is still reliant upon semaphore stations for long-distance communication. The idea of a Victorian world which features anachronistic inventions such as computers or televisions is not a modern conceit, but was a very popular one in the Victorian period itself. Illustration 10 shows us a middle-aged couple in London, Skyping with their daughter who has been playing tennis in Sri Lanka. They are using for this purpose a very large wall-mounted, flat-screen television. This scene appeared in the magazine *Punch* in December 1878.

In Chapter 1 we explored the uncanny similarities between nineteenth century predictions of this kind and the sort of scenarios which we routinely encounter in the steampunk genre. The cartoon from *Punch*, drawn incidentally by George du Maurier, grandfather of the more famous Daphne, was not atypical of the period. Other artists and writers were also speculating on what wonders might be just around the corner and the idea of television exercised a particular fascination, which was not limited to fictional representations.

On Thursday, 15 January 1885, the Imperial Patent Office in Berlin granted to Paul Gottlieb Nipkow a patent for an 'electric telescope', this being for the 'electric reproduction of illuminated objects'. It was, in other words, a patent for television. Because television programmes these days are nearly all carried by digital signals and the picture formed from LEDs or plasma by complicated electronic means, we sometimes forget that there are simpler ways of building television sets. Only a few years ago, cathode ray tubes were universally used for the receiving of television transmissions. These used rays of electrons which were projected onto a phosphorescent screen. Before that though, some 80 years ago, there was mechanical television. This was the system which John Logie Baird perfected in the 1920s and which for a few years was seen as the best and most efficient system.

Following the invention of the telephone in the 1870s and the realization that sound could be turned into electrical impulses and sent through a wire, it was only logical that some people would start to wonder if the same thing could not be done with light. Fax machines were known and when in 1873 two English telegraph engineers discovered that a selenium wire changed its conductivity according to the amount of light falling upon it, the way seemed open to producing a machine analogous to the telephone, but which would transmit images, rather than sound, along a wire. After Alexander Graham Bell invented his photophone, which actually turned light into electrical signals capable of carrying articulate speech, it seemed to be only a matter of time before something like television appeared.

As early as 1880, an article was published in the French magazine *La Lumière électrique*, setting out the five features which any practical television system would need. According to the author of the piece, Maurice Leblanc, these were as follows. Firstly, a transducer would be needed to convert light into electricity. Such a photo-electric cell was already being used, of course, in Bell's photophone. Secondly, a scanner would be needed to break the image down into pieces. A method would be needed for synchronizing this scanner with a similar device at the other end of the line. The two other things necessary would be a way of turning the electrical signals back into light and also a screen on which to display them.

It is worth mentioning that these five points are all as applicable today as they were in 1880. Modern television broadcasting follows a set of parameters which were first set out almost 140 years ago.

The first step in the construction of a practical television system was taken by Paul Nipkow, with his 'electric telescope'. What the German engineer drew up plans for was a rotating disc with holes punched in it in a spiral pattern. This was to be set spinning in front of a selenium cell and would have the effect of breaking down a scene into separate and distinct parts. The light passing through each hole would strike the selenium and, according to its intensity, vary the amount of electricity which was passed through the cell. This signal would then be sent to a receiver; another spinning disc placed in front of a gas discharge light. The two discs would be synchronized. As each hole passed in front of the lamp, the level of light seen would correspond exactly with the amount of light falling onto the selenium cell through the hole in the transmitting disc.

Nipkow's proposed television transmitting and receiving apparatus was perfectly sound; it was after all the very method which John Logie Baird used in the 1920s. Baird, whose system was used in the first BBC television broadcasts before the Second World War, used a Nipkow disc to break down and reassemble the images which were sent by radio waves. As a matter of fact, detailed plans for building a television, both transmitter and receiver, were published in America in the same year that Nipkow's patent was granted in Germany. The 14 November 1885 edition of the magazine *Electrical World* contained an article headed, 'Seeing by Electricity'. This explained the nature of Paul Nipkow's invention and explored the implications.

The piece carried by *Electrical World* is really quite staggering to read. This is because it not only describes in detail how television could work, using Nipkow's method, which of course John Logie Baird discovered in the 1920s, but it also talks casually of infra-red photography and television cameras which could see in the dark. Like Alexander Graham Bell's photophone, Nipkow thought that a carbon-coated surface would need to be used in conjunction with the selenium cell. As *Electrical World* explained;

> Herr Nipkow points out that by means of this apparatus objects in the dark can also be seen, provided they are sufficiently heated, since the lamp-black drum also responds to the non-visible rays of the spectrum.

Incredible as it may seem, the principles of infra-red television cameras were being expounded almost 135 years ago.

It is not known if Paul Nipkow actually constructed a transmitter and receiver based upon the patent which he was granted in 1885. Certainly, 15 years later, when it came up for renewal, he allowed the patent to lapse. This suggests that whatever experiments he carried out had not shown the system to be viable using the technology of the time.

The late nineteenth century was a time when television, like radio, was constantly on the horizon. It was talked of and written about extensively and even the readers of humorous magazines such as *Punch* were aware that it was a matter of time before they saw it in their homes. The word 'television' itself was being used before the death of Queen Victoria. On 24 August 1900, Constantin Perskyi read a paper to the International Electricity Congress, which was being held as part of the International World Fair in Paris. In his speech, Perskyi mentioned television and talked of the practicalities of the thing. There were those who objected to the word and regarded it as being illegitimate, being composed as it is of a mixture of Latin and Greek. The 'tele' part of the word is from the ancient Greek for 'far', while 'visio' is Latin for 'I see'.

Thinking of television brings us to one of the strangest uses of technology in nineteenth century Europe and America. We are today quite familiar with the idea of cable television, that is to say television programmes which, rather than being transmitted through the air on radio waves, are sent along wires to the consumer's home. Our pavements have in the last few decades been dug up to provide this service to whole streets at a time and 95 per cent of the United Kingdom is now able to be connected to this system. Closed-circuit programmes of this kind are nothing new. Queen Victoria was a subscriber to a very similar system which was popular as the century drew to a close.

The names of some inventors crop up again and again in wildly differing fields. Hiram Maxim, in addition to building and taking off in Britain's first aeroplane, was also one of the men who made the petrol-driven car a practical proposition by devising a way of suppressing the awful noise which the early motor cars made. Another pioneer of powered flight was of course Clement Ader who, in 1890, made the first unassisted take-off in a powered, heavier-than-air flying machine. Ader, in addition to his skills as a mechanical engineer, was also interested in electrical telecommunications.

The first stereophonic gramophone records did not appear until 1958 and at roughly the same time FM radio broadcasts began in stereo. For the public in both Europe and the United States, music in stereo like this was a great novelty. It was not, however, the first breakthrough in transmitting music in stereo. That occurred in 1881 and was the invention of Clement Ader.

At the International Exposition of Electricity, which was held in Paris between 15 August and 15 November 1881, an amazing demonstration was held of the future possibilities of the telephone, which had been invented just five years earlier by Alexander Graham Bell. Clement Ader, who had invented the new system on show at the exposition, had arranged for microphones to be placed at the front of the stage of the Paris Opera. Cables then ran to the hall of the exposition, two kilometres away. In a special room, visitors could listen, with an earpiece in each ear, to a stereophonic performance. This was sensational and a number of well-known people expressed their approval. Victor Hugo, the author of *Les Misérables*, was very impressed. The future for a commercial service based on Ader's idea looked very bright.

It was some years before the technical problems could be overcome, but in 1890 the Theaterophone live music service was launched in Paris. Before this, various experimental lines had been installed in Portugal, Belgium and Sweden. There certainly seemed to be a demand for hearing musical performances when people were unable or unwilling to go to an opera house or theatre in person. The cost of having a Theaterophone line on your home was prohibitive to any but the very well-off, however. The initial outlay was 180 francs for the equipment and then another 15 francs for each performance listened to. To give some idea of how much this might be in modern terms, at that time it was possible to rent a flat in a good area of Paris for perhaps 60 francs a month. For the less prosperous, Theaterophone booths were set up in cafes and other public places, where one could listen in for a mere 50 centimes.

Of course, orchestras were not playing 24 hours a day and so something had to fill in the blank periods. This was done by news bulletins, weather forecasts and music from an automatic player-piano. It did not seem to matter to those parting with their 50 centimes what was being transmitted; it was the novelty of the thing which was attractive. In 1892, a similar system was demonstrated at London's Crystal Palace and two years later the Electrophone service began. For £10 a year, a substantial sum at that time, a receiver could be rented and a selection of theatrical performances heard. Later, churches too were wired into the network and subscribers could save themselves the trouble of attending divine worship in person. At its height, Electrophone had 2,000 people hiring its equipment, including Queen Victoria.

There was no doubt that home stereo, a kind of prototype Spotify, was generating a great deal of interest in the 1890s. The cost was too high for most people and so the race was on to make it something which was affordable to ordinary consumers. This was most nearly achieved in Hungary, where the Telefon Hirmando telephone newspaper began in 1893.

At its height there were 15,000 subscribers to Telefon Hirmando in Budapest; capital city of Hungary. The whole service was eerily reminiscent of later radio schedules and it is possible that early broadcasters actually copied the Telefon Hirmando format. Here is how the morning timetable began

   9:00   Precise time signal
   9:30   Foreign news
   10:00 Stock Exchange quotations
   10:30 Roundup of local news headlines
   11:00 General news and finance

Microphones were permanently installed in the Budapest Opera House, so that listeners could relax in the evening and not have to go trekking off to hear the opera in person. This situation prompted a writer in the magazine *Harper's Weekly* to remark that the service had made Budapest the finest city in the world for 'the illiterate, blind, bedridden and incurably lazy people'. Precisely the same kind of observation being made today about Facebook and YouTube addicts.

By 1901, over 5,000km of wire hung above the streets of Budapest, carrying the Telefon Hirmando news and music into 6,200 homes. It seemed that telephone newspapers like this and Electrophone were truly the thing of the future. Of course, within a decade or two, the boom had ended, killed by radio. Why pay for an expensive, subscription service when for the cost of a radio set, you could listen for nothing to any programmes you wished?

We will finish by looking at something which almost everybody in the developed world uses, directly or indirectly, on a regular basis. Hard drives are spinning discs, usually of metal, upon which information is stored by magnetic impulses. Many laptops and most larger computers use this technology to record information permanently. It seems almost beyond belief that this method of storage was invented and in use before the death of Queen Victoria, but such is the case.

Few people today are likely to have heard of Valdemar Poulsen, who was at one time known as the 'Danish Edison'. Poulsen was working in the 1890s for the Copenhagen Telephone Company and one day had a curious idea. The very operation of the telephone entailed converting the sound of the human voice into electrical signals and sending these along a wire. How if those electrical signals could be used to magnetise a wire and thus store them in permanent form? This was at a time when the only means of recording the human voice were crude phonographs and gramophones, which relied upon engraving grooves in wax, shellac or tinfoil.

The experiment which Valdemar Poulsen devised to test his idea was simplicity itself. He stretched a length of steel piano wire at a gentle angle from one end of a room to the other. Then he attached to it an electromagnet, so that it could slide down the wire freely, under the force of gravity. To the electromagnet, Poulsen connected a microphone and, as the electromagnet slid down the wire, he spoke into the microphone. After doing this, he repeated the process, only this time connecting a telephone earpiece to the electromagnet, rather than the microphone. Listening through the earpiece, Poulsen was enchanted to hear the sound of his own voice. He had invented magnetic recording. Actually, he had done a great deal more than that.

In a patent application, Poulsen outlines some possible developments of his discovery;

> . . . as a receiving device a steel band, supported if necessary on an insulating material and brought under the action of an electromagnet. Such an arrangement has the advantage that a steel band of a desired length may be used. Instead of a cylinder there may be used a disk of magnetisable material over which the electromagnet may be conducted spirally; or a sheet or strip of some insulating material such as paper may be covered with a magnetisable metallic dust and may be used as the magnetisable surface. With the aid of such a strip which may be folded, a message received at any place provided with the new apparatus may be sent to another place where it may be repeated by passing the strip through the apparatus at that place.

Valdemar Poulsen had, in the 1890s, laid the groundwork for tape recorders, telephone answering machines, magnetic strips on bank cards, floppy discs, hard drives and much else besides. By showing that it was possible to store information permanently as magnetic domains on metal, he had made the computer revolution of the late twentieth century possible.

An amusing by-product of the Danish engineer's simple experiment with a piece of piano wire was that he invented the telephone answering machine at a time when there was no call at all for such thing. Poulsen called his recording machine using steel wire the 'Telegraphone' and when he demonstrated it at the World Exposition in Paris in 1900, it was a great hit. Linked up to a telephone, it enabled a caller to leave a message, even when the person being called was out. What was the point of that though? The telephone answering machine was simply an expensive novelty, of no practical use. One or two offices installed them, but it was to be another 60 or 70 years before they really caught on. One place in England which did buy a telephone answering machine was the Royal Dockyards. The one which they used had a hard drive to store messages; a large metal platter which revolved past a recording head.

We have looked in this chapter at various inventions which are often regarded as being quintessentially of the twentieth century; the optical transmission of telephone calls, fax machines, hard drives, television and answering machines. To our surprise, we have learned that they were all invented during the reign of Queen Victoria; decades before the beginning of the twentieth century. Before ending this chapter, we will look again at a subject touched upon in Chapter 1; that of Victorian robots, fictional and real.

As we have seen, Boilerplate, the imaginary steam-powered robot created at the beginning of the twenty-first century, is based upon ideas from the nineteenth century. Some of these are to be found in lurid dime novels and others in patent applications and newspaper articles. So ingenious have been the creators of Boilerplate, that their imaginary robot is taken by many people on the Internet for true history and as a consequence the dime novels, real Victorian inventions and the modern steampunk fantasy all become jumbled up, one with the other, until it is all but impossible to distinguish fact from fancy.

Steam, and later electricity, were the motive forces behind most transport at the end of the Victorian period. Steam trains, steamships, electric trams, electric cars, steam wagons and steam aeroplanes; these represented and symbolized the future. Making a steam-powered or electrically-driven metal man or horse was just a natural extension of this trend. Despite the elaborate and detailed technical drawings which accompanied the applications for patents for mechanical men or horses, there is little evidence that any of them were able to do much, apart from move their legs up and down in a rough facsimile of natural movement. Most were little more than dummies.

When steam trams were introduced to California in the early 1870s, they caused a great deal of disturbance, due to the panicky reactions of horses to the sight of a steam-powered vehicle. Horses bolted and if they were harnessed to a cart, then the results could be serious. In 1876 an inventor called S.R. Mathewson patented his own solution to the crisis brought about by the trams. His answer was a dummy horse which contained the steam engine. This would, he hoped deceive any horses which were encountered into mistaking his model for the real thing. Although a prototype was trialled, it was not a notable success.

Inspired perhaps by Mathewson's strange tram, some inventors turned their minds to designing mechanical horses which could actually walk, run and be programmed to do useful work, such as pulling carriages or hauling ploughs. Despite persuading investors to put their money into such futuristic schemes, none of them ever produced a working model. Claims were often made years after the event that this or that model had worked, but the evidence is not convincing.

While on the subject of steam-powered animals, it will come as no surprise to learn that Jules Verne's inventions knocked everybody else's into a cocked hat. Others might dream of mechanical horses, but Verne wrote the tale of a steam-powered elephant. In 1880, a novel called *La Maison à Vapeur* (*The Steam House*) was published. Set in the aftermath of India's Sepoy rebellion of 1857, it tells the story of a great wheeled house, like the most luxurious of caravans, which is drawn along by a steam-powered robot elephant. A group of British travellers use the eponymous steam house to cross India, enjoying many adventures on the way.

Chapter 7

The Quest for Renewable Energy

A news item in May 2015 captured the imagination of those worried about global warming and our dependence on the burning of fossil fuels to generate electricity. A research station in South Africa's Kalahari Desert revealed how electricity was being produced by 'zero emission' generators, powered by solar energy. It looked a very exciting project, one perfectly in keeping with modern times and likely to play to the anxieties of those worried about climate change and pollution. The accompanying photographs, showing parabolic reflectors in the desert aimed at the sun, looked like something from a science fiction film. This could certainly be the future of 'clean' energy. Except of course, that it was nothing of the sort. The technology being used at this pioneering pilot-plant was rooted firmly in the nineteenth century.

In Illustration 9, we see a solar power plant. The juxtaposition of the curved reflectors, which look a little like some sort of death-ray, and the people wearing top hats and ankle-length dresses is disconcerting. This is not a modern-day drawing, though, from a graphic novel. It is an engraving of a scene at the Paris Exposition in 1878, when Auguste Mouchout set up a solar power plant and used it to make, of all unlikely things, ice cubes. A similar device may be seen in Illustration 5, where the sun's rays are being focused and used to drive a printing press. The technique used in these two contraptions is identical to that which was to be used in the supposedly revolutionary power plant being set up in the Kalahari Desert.

The parabolic reflectors are one part of the project in the Kalahari Desert which was seemingly so much in tune with the spirit of our age, which demands clean and renewable energy, rather than that obtained by the burning of fossil fuels. The 'zero emission' generators, which would be powered by the heat of the sun, were nothing more than Stirling Engines. The Stirling Engine is a device invented by a Scottish clergyman called Robert Stirling in 1816, just a year after the Battle of Waterloo. Rather than being a ground-breaking and cutting-edge instance of modern technology at work, the experimental plant in South Africa had really just cobbled together two nineteenth-century gadgets.

It is today regarded as axiomatic in the Western World that Europe and America must reduce their dependence on fossil fuels of all kinds; the routine use of oil, coal and natural gas must become a thing of the past. There are two chief reasons for this desire to free ourselves from the need to extract carbon-rich fuels from the Earth. One of course is that their use contributes to climate change. The other is that fossil fuels are in limited supply and when they are gone, they are gone. Perhaps not this year, maybe not even this century, the oil that we use for our cars will run out and we will have mined or obtained by fracking all the remaining coal and gas. If we have not by that time found alternative ways to generate electricity, then there is the prospect of our civilization entering a new Dark Age, with industrialization coming to an abrupt halt. This is a chilling scenario and one which would have been very familiar to the Victorians.

Efforts in our own time to reduce dependence on fossil fuels have resulted in the appearance of solar panels on rooftops, wind turbines which generate electricity and various other schemes to produce power without burning oil, gas or coal which we take out of the ground. In doing so, we are following in the footsteps of various nineteenth-century pioneers, who laid the groundwork for the technology which we are today exploiting in this field. Before looking at some of the inventions which, it was hoped, would free the Victorians from the need for drilling and mining for carbon-rich deposits in the earth, let's see what the specific fear was which drove such developments over a hundred years ago.

On Thursday, 1 September 1881 the famous scientist Sir William Thompson, later to become Lord Kelvin, after whom the Kelvin temperature scale is named, gave a speech in Edinburgh to the British Association for the Advancement of Science. His tone was sombre and Lord Kelvin urged others to share his anxiety about what he saw as a looming crisis. The title of the paper he delivered that day was *On the Sources of Energy in Nature Available to Man for the Production of Mechanical Effect*.

Sir William wanted his listeners that day to understand one very important point; that their entire civilization was founded upon the burning of fossil fuels and that when those fuels ran out, they were likely to be in terrible trouble. It was coal which had fuelled the Industrial Revolution and it was still powering British society at the time the talk was delivered. It was burned to power the railway trains which spanned Britain, it enabled ships to cross the ocean, it was vital for the production of steel and it was used in the generation of electricity. Lord Kelvin though, feared that it was running out. He told the audience that day;

The subterranean coal-stores of the world are becoming exhausted surely, and not slowly, and the price of coal is upward bound—upward bound on the whole, though no doubt it will have its ups and downs in the future as it has had in the past, and as must be the case in respect to every marketable commodity. When the coal is all burned; or, long before it is all burned, when there is so little of it left and the coal-mines from which that little is to be excavated are so distant and deep and hot that its price to the consumer is greatly higher than at present, it is most probable that windmills or wind-motors in some form will again be in the ascendant, and that wind will do man's mechanical work on land at least in proportion comparable to its present doing of work at sea.

It was, said Sir William, imperative that new ways of producing power were developed. He spoke in detail of solar energy, wind turbines and tidal booms, the very things in fact that we are currently working on to produce electricity so that we too may break our reliance on fossil fuels.

Periodically over the last century or so, fears have been expressed about dwindling reserves of oil in the ground, which, when they run out, will leave our civilization in a desperate state. During the 1970s, such anxieties reached a peak, with predictions being made that oil would run out within 30 years. The fears about coal, which from time to time gripped nineteenth-century scientists, were very similar. As early as 1829, the British government summoned leading geologists of the day to London, so that they could be quizzed about the likelihood of the country's coal deposits running out. At the dawn of the twentieth century, 70 years later, such fears were still being expressed. In 1903 *Harper's Weekly* observed that;

It has been a favourite pastime for the dreary gentlemen who juggle with statistics, solemnly to calculate the date on which we shall all freeze to death from exhaustion of the coal supply.

Auguste Mouchout, the man who devised the first practical solar boiler, which can be seen in Illustration 9, was also concerned about the prospect of European countries running out of coal and seeing their industries brought to a grinding halt. Like Lord Kelvin in Britain, Mouchout tried to alert people to the supposed peril, saying, 'Eventually industry will no longer find in Europe the resources to satisfy its prodigious expansion. Coal will undoubtedly be used up. What will industry do then?'

The fear of running out of coal, unrealistic though it was, spurred on the invention of many very modern-looking devices. Some of the machinery which was being produced at that time has a steampunk feel about it. They appear both startlingly modern and also weirdly anachronistic. Modern, because they are connected to parabolic mirrors, which puts us in mind of satellite dishes, and anachronistic because they are being run by a steam engine. A few words may not come amiss at this point about the generation of electricity.

We often use the word 'steam' to signify that something is hopelessly outdated and old-fashioned; 'steam radio', for example. The steam age is understood to be something in the past. We see this use of the word 'steam' in the very title of this book. Steampunk is obviously not about our modern world, simply because it does feature the word 'steam'. In fact, almost all the radios and televisions that we listen to and watch today are, in reality, driven by steam. Steam engines produce the great majority of power in Britain today; we are living in the Age of Steam. Only by understanding this will we be able to see what was happening in Victorian Britain and appreciate their anxieties about the depletion of natural resources.

To make most of the electricity in this country, a source of heat is used to boil water and so produce steam. Sometimes coal or gas are burned, but allowing a mass of uranium to get very hot can produce the same effect. Once the water is boiling and steam is being generated, then this is directed through turbines and makes them spin. The rotary movement turns generators and so gives us electricity. Steam power is an integral part of electricity generation in the modern world.

The way that we generate electricity today, by means of rotating, steam-driven turbines, was established in the nineteenth century and is now so common that we regard any other means of producing electricity freakishly novel. This is a wholly mistaken view. When we install arrays of photovoltaic panels on rooftops or talk of using wind and water to produce electricity in a renewable way, we are actually turning back to Victorian technology, rather than looking forwards towards modern ingenuity. One or two examples might make this clearer.

The first use of electricity to produce light was the arc light, where a continuous bright spark was produced between two terminals. This is all right for floodlights or street lamps, but too harsh for domestic use. Homes were actually being illuminated using arc lamps before 1880, but they were never likely to catch on. Colonel R.E.B. Crompton, the engineer, had arc lights in his house at Porchester Gardens in London in December 1879.

These were powered by giant batteries; another modern notion when it comes to domestic use of electricity. Even earlier that year, there were reports that a Mr H.W. Tyler was having arc lights rigged up indoors at his country seat of Wyvenhoe Hall, near Colchester in Essex. These too were powered by batteries. At Cragside, a grand country house in Northumberland, arc lights were also being used as early as 1879, but were abandoned and replaced with state-of-the-art incandescent bulbs, the type of electric lamps which would be in use for the next 130 years or so. These were powered not by inefficient batteries but rather by a hydroelectric plant, which even then seemed to some to be the way forward for the generation of electricity.

William Armstrong was a typical Victorian captain of industry. Like Hiram Maxim, Armstrong made his fortune from the armaments business, being widely regarded as the father of modern artillery. Once he had become rich through his designs of long-range weaponry, William Armstrong decided to have a grand house built in a fairly remote country area and to live a life of leisure. It is not always easy to stop inventing things though, once you are in the habit of doing so, and in 1878 Armstrong came up with a scheme for making his house the most modern, in one way, in the whole country. He decided that the whole place should be illuminated by electricity, using not the wasteful and inefficient arc lights that other people favoured, but rather the incandescent light bulbs which had recently been perfected by Joseph Swan in Britain and Thomas Edison in America.

Electric light bulbs which needed no adjustment and would burn steadily for many hours were a revolutionary idea when William Armstrong chose them to light his home. So too was the scheme by which he intended to produce the electricity to power all those lights. A stream ran through the land surrounding Armstrong's house, which he had named Cragside. By arranging for dams to be built, reservoirs of water were used to power waterwheels, which turned dynamos and generated electricity. The whole system was what we today describe as being 'carbon neutral' or 'environmentally friendly'.

We come now to one of those points of divergence in history. Readers will probably be aware that most steampunk narratives require a point of divergence, where some development causes history to change and become unrecognizable. This might be an invention, or it might be losing rather than winning a war, or perhaps the premature death of somebody who grew, in the real world, to become immensely influential and important; the possibilities are limitless. In our own world, from 1882 onwards electricity began to be generated almost exclusively by the burning of coal, which is a fossil fuel. We are today trying to move away from this dependence upon fossil fuels to make electricity. It is for this reason that wind turbines, solar power and hydroelectric plants are being exploited.

In 1881 Sir William Thompson had lectured that dependence upon fossil fuels was a terrible mistake and that it would be better to rely upon renewable energy such as the power of the wind, instead of simply burning coal to produce steam. Such views were taken seriously, because in that same year the world's first public supply of electricity began, which provided both street lighting and power to ordinary homes. Instead of being produced by burning fossil fuels, the electricity was generated by environmentally-friendly means.

The quiet English country town of Godalming might seem an odd place to find the world's first provision of a public electricity supply, available both for lighting the streets and for domestic use. One would perhaps have expected to hear of such an exciting scheme first being instituted in New York or London, rather than a backwater like Godalming in Surrey. Nevertheless, the Saturday, 1 October 1881 edition of the local newspaper, the *Surrey Advertiser*, carried the following story;

> On Monday evening (the 26th September) the upper portion of the Borough of Godalming was lighted by electricity for a few hours as an experiment and continued each night since, the motive power to generate the current being an auxiliary face water-wheel at the Westbrook Mills, of Messrs. Pullman brothers, the skin dressers, who have made arrangements for lighting their mills with the Swan lights, and for the larger open spaces with Siemen's differential lamps of 300 candle-power each.

This article is chiefly concerned with the public exploitation of the new electrical supply, but it was also available to ordinary houses as well, if the owners could afford to pay for it. Electric lighting worked out a little more expensive at that time than gas.

First one house had been supplied with electricity generated by flowing water and now an entire town was similarly given a public supply using the same method. It was now touch and go whether this was to be the great thing of the future. As many scientists pointed out at the time, mining coal in order to burn it, this heating water and producing steam to run a steam engine which would turn a dynamo, was a very roundabout and inefficient way of achieving things. It made far more sense simply to harness the power of rivers and tides to turn the dynamos directly, without using up any coal at all. This tied in neatly with fears about the depletion of coal stocks and the anxiety that they might soon be exhausted. The world was on the cusp of a clean energy revolution, one which would not rely upon fossil fuels.

Not everybody could see the great opportunity which presented itself with the successful exploitation of hydroelectric power. For many people, steam was modern; it was the driving force of the Industrial Revolution and the obvious choice when setting up a power station. There was, it was felt by some pioneers in the field, something a little backward-looking and old-fashioned about the idea of using water wheels to make electricity. Steam engines were modern; they were *chic*. There was plenty of coal, why not exploit it to the full?

Three and a half months after the world's first supply of public and domestic electricity came online in Surrey, another power station opened in the heart of London, at Holborn Viaduct. This was built by the Edison Electric Light Company and turned out 93kw of DC electricity at 110v. It was driven by a coal-fired steam turbine. Thomas Edison was perhaps the most influential figure in electrical engineering at that time and if he had plumped for steam turbines to turn his dynamos, then that was good enough for everybody else. From that day to this, the burning of fossil fuels to make steam to turn turbines has always been the most popular way of producing electricity, not only in this country but throughout the entire world. Hydroelectric power became a minor and insignificant player in the field. Just as clean, hydroelectric power stations almost became the standard, so too did solar energy enjoy a brief vogue at roughly the same time.

There are two ways of using the energy of the sun to perform work. One is by harnessing the sun's rays and using their heat to perform work. The other is to convert the light directly into electricity by the use of photovoltaic cells. Both have their roots in the nineteenth century. Glass-covered boxes which are painted black can heat up very rapidly when the sun shines on them. These are often to be seen on roofs in Mediterranean countries, where they provide hot water. Such an arrangement can also be used to cook food, as astronomer Sir John Herschel found when he visited South Africa in 1834. By that time, the technology of solar cooking was well-known.

Once again, it was fears about the wisdom of relying upon fossil fuels which spurred on the development of alternative sources of energy. Just as with hydroelectric power, there came a point at which the whole idea so very nearly took off, before short-sighted political and economic considerations intervened and stifled common sense.

In nineteenth-century France, some people were voicing the same fears as Lord Kelvin in England; that supplies of coal would run out and the

Industrial Revolution grind to a halt. It was a terrifying prospect and since the price of coal in France was, for various reasons, rising at that time, the search began for other ways of obtaining energy. Augustin Mouchot was a French mathematician and teacher, who was born in 1825. He was also an inventor and from 1860 devoted his efforts to finding alternative energy sources and so free his country from reliance upon coal, much of which was being imported from abroad.

To begin with, Mouchot experimented with solar cooking and found that it was perfectly possible to cook a meal without burning any wood or coal, simply by enclosing a black container in glass. The greenhouse effect did the rest. This was interesting, but hardly new. His next experiment though was certainly exciting and novel. Mouchot managed to operate a small steam engine in the same way; by solar power alone. He did this by concentrating the sun's rays by a parabolic mirror. One of his larger mirrors of this sort may be seen in Illustration 9. The emperor of France, Napoleon III was very interested in Mouchot's work, seeing the economic benefits if France's use of coal could somehow be reduced.

In 1869 Augustine Mouchot set out his theories about solar energy in a book called *La Chaleur solaire et ses Applications industrielles*, which translates roughly as 'Solar heat and its Industrial Applications'. The Franco-Prussian War, which began the following year, disrupted Mouchot's work, but by the following year he had gained permission to install a solar generator at the library in Tours. A few years later, he was given leave of absence from his teaching post so that he could construct his biggest solar powered engine yet for the Universal Exhibition of 1878. It is this which may be seen in Illustration 9. The engine, powered only by focused sunlight, ran a small refrigeration plant which, to the amazement and delight of visitors to the exhibition, produced a steady supply of ice cubes. Before the Universal Exhibition in Paris though, Mouchot had been given facilities to travel to Algeria, then a French colony, where the almost constant supply of bright and uninterrupted sunshine made it an ideal place to conduct experiments in solar energy. Mouchot set up desalination plants in Algeria, which turned seawater into drinking water.

The French government, with the approval of Napoleon III had, up until this time, been financing Mouchot's work. It was thought that if solar energy could help to reduce France's reliance on imported coal, then it was well worth encouraging. As is so often the case with governments though, the only real interest lay in short-term advantage, rather than any vision of a future world which might be improved. As soon as the immediate need for such research disappeared, the funding was likely to be axed at once.

On 23 January 1860, Britain and France signed what has been described as the world's first modern trade agreement. The Cobden-Chevalier Treaty was a free trade agreement, which removed or reduced tariffs and duties on goods traded between Britain and France. Because coke and coal were specifically included in the Cobden-Chevalier Treaty, this had the effect of gradually bringing down the price of coal in France. Combined with improvements in transport, which allowed coal to be moved about the country more easily, the French government decided at the end of 1878 that they no longer needed to worry their heads about alternative energy. All funding was cut for projects such as Mouchot's and France's brief flirtation with solar power came to an abrupt end.

In America, interest in using solar energy to drive machinery was being expressed by various scientists, including one man of whom we have already read. Samuel Pierpoint Langley, of the Smithsonian Institution, was to become famous at the end of the nineteenth century for his efforts in the field of powered flight. In 1884 though, another topic was exercising his mind. Before we see what he was writing about in that year, a slight detour will be necessary, back to the year following the Battle of Waterloo. In 1816 a Scottish clergyman called Robert Stirling invented and patented a new type of engine. The Stirling Engine, as it came to be known, worked on a completely closed cycle of either air or certain liquids. Unlike steam engines, there was no need for a constant supply of water; the Stirling Engine was self-contained. All it required to operate was a source of external heat, to be applied to one end of the engine.

The internal details of the Stirling Engine need not concern us here. It is enough to know that by applying heat to one part of it, a cylinder is caused to move back and forth, which in turn operates a crank and produces rotary movement. There have been many improvements to the Stirling Engine over the two centuries since its invention. One man who improved upon the Stirling Engine to such an extent that he claimed that his work on the idea amounted in effect to a completely new invention was an American called John Ericsson. Ericsson was originally from Sweden and spent his life inventing things. He built a steam locomotive called *Novelty*, at the same time that George Stephenson was working on his *Rocket*. Moving to America, John Ericsson invented the rotating gun turret for warships, which was first installed on the USS *Monitor* and gave the Union fleet a distinct advantage over that of the Confederates during the American Civil War.

One of John Ericsson's main concerns was the same thing which was troubling a lot of scientists and inventors as the nineteenth century drew to a close. He thought that the Middle East would become very important from the point of view of energy supplies, not because of the oil which is there, but rather due to the abundant supply of sunlight. He wrote;

> The rapid exhaustion of the European coal-fields will soon cause great changes in reference to international relations, in favour of those countries which are in possession of continuous sun-power.

Ericsson had followed Auguste Mouchot's work with interest and spotted one great flaw in what the Frenchman had been doing. It takes a great deal of energy to heat water to boiling point and using the sun's heat in that way to boil water and then use the steam to operate the energy was a wasteful way of going about things. Far better and more effective if the heat could be made to run an engine without the intermediate stage of supplying water and having to adjust pressure valves and tinker with an engine to ensure that it had a supply of water and so on. What John Ericsson did was to refine and redesign the Stirling Engine until it was as efficient as could be. Then he placed it at the focal point of a parabolic reflector and allowed the rays of the sun to do the rest. The result was an engine which would turn without anybody having to do anything at all to keep it working smoothly.

In December 1884 an American magazine carried something likely to excite readers greatly; an extract from Mark Twain's latest book, *The Adventures of Huckleberry Finn*. Also in that same edition of *Century Magazine* was an article by Samuel Pierpoint Langley about solar energy. Langley was a famous scientist at that time and like many in the scientific world, he was concerned about the consequences for the world of running out of fossil fuels. He also wanted to draw attention to the fact that burning fossil fuels, even when they were plentiful, polluted the atmosphere as waste heat. All of which has a decidedly modern flavour about it. Since the fossil fuels ultimately contained energy which had been absorbed from the sun's rays, why not skip a stage in the process and utilize the rays directly?

Langley proposed that arrays of parabolic reflectors, with Stirling Engines attached, could be set up in the Sahara Desert, where they would provide limitless free energy which would allow the desert to be irrigated and colonized. It will be noticed that this is precisely the same as the radical scheme which was mentioned at the beginning of this chapter, which involved parabolic reflectors linked to Stirling Engines and being set up in another African desert, the Kalahari. Far from being an up-to-date and modern way of harvesting solar power, this was no more than a straight copy of a scheme first suggested over 135 years ago.

In exactly the same horribly short-sighted way that the French government abandoned any idea of developing energy sources that were not reliant upon fossil fuels, as soon as those fuels were abundant and cheap enough, so too the United States. True, with a little research it would have been perfectly possible to create solar-powered engines and even use them for generating electricity, but with the seemingly unlimited supplies of oil and coal which America had in the late nineteenth century, why bother? If you wish to run an engine, it was much easier to power it with petrol, rather than fiddle about with reflectors to gather sunlight.

Private enterprise almost brought about the revolution in energy production and just before the outbreak of war in 1914, it really looked for a time as though the world would soon be getting most of its power from solar energy. An American called Frank Shuman began experimenting in 1906 with a glass-covered black box, which he found had the ability to trap the sun's heat and use it to heat water. The following year, he covered the whole of the back garden of his home in Philadelphia with a bewildering array of black pipes, which he had filled with ether. This has a low boiling point and was ideal for Shuman's purposes. The 96m² generated 2.2 kilowatts and was used to run a water pump. In winter, the amount of work done by this solar-powered system dropped, but it was still considerable.

The Sun Power Company was founded by Frank Shuman, following his successful experiments and he found no shortage of people willing to invest. In 1912, he moved to Egypt and set up the biggest solar power plant that the world had ever seen. It was established a few miles from Cairo and covered 1,200m². The mirrors and pipes of the installation at Meadi produced 40 kilowatts and ran 24 hours a day. Even the Consul-General of Egypt, Lord Kitchener, came to visit Meadi and was most impressed by what he saw there. He offered Shuman a plot of land south of Egypt in the Sudan. This would enable him to set up a plant ten times as big as that in Meadi.

All the indications were that solar power was now the big thing of the future and Frank Shuman began to plan for the ultimate solar power plant, one which would provide enough energy for the entire human race. He set out his ideas in the *Scientific American*. The figures were fairly straightforward. All the oil and coal taken from the earth in 1909 produced about 270 million horsepower, or 200 gigawatts in modern terms. If 20,245 square miles of the Sahara Desert were to be covered in one of his arrays, it would yield this much energy. With unbounded optimism, he ended by suggesting that, 'Surely from this showing, the human race can see that solar power can take care of them for all time to come.'

The article in *Scientific American* appeared in the February 1914 edition. Before the year was ended, the world had far more urgent matters to attend to than a new scheme for producing cheap energy. As for Lord Kitchener, he found that he was a little too busy in 1914 to devote much time to helping set up a power plant in Sudan. By the time the First World War had come to an end, the economies of the world's industrialized nations were in ruins and nobody was inclined to invest heavily in solar power. Oil was cheap and there was still plenty of coal.

We end this examination of clean and carbon-neutral energy production by thinking of that most modern phenomenon, the photovoltaic roof panels which we increasingly see on rooftops in Europe and America. These are a very recent addition to city streetscapes and it is easy to jump to the conclusion that solar cells are relatively new. Once again, this is a wholly mistaken perspective. The technology for generating electricity from sunlight dates, like so many other apparently modern inventions, from the nineteenth century.

The ways at which we have so far looked for exploiting the sun's energy have all worked indirectly, by using the heat of the sun to run steam engines or self-contained engines such as the Stirling Engine. The rotary motion thus obtained is then used to carry out useful work; driving a pump, for instance, or turning a dynamo and generating electricity. If one wishes to produce electricity then there could hardly be a more roundabout way of going about it. The truth is that given the right conditions, sunlight can be converted directly into electricity.

In 1839 French scientist Edmond Becquerel was tinkering with an electrolytic cell, one which produces a chemical reaction when electricity is passed through it. This can be something like the electrolysis of water, breaking it down into its component elements of hydrogen and oxygen. Becquerel noticed something very strange. In one of his experiments, he had a solution of silver chloride in acid, with two platinum electrodes immersed in the solution. When light was shone on this arrangement, electricity was generated from the electrodes. Unwittingly, Edmond Becquerel had invented the world's first solar cell.

The amount of current produced by Becquerel's cell was tiny and it was not enough to perform any useful work. In 1876 William Adams and Richard Day showed that when selenium was exposed to light, a current was generated. This was surprising and counter-intuitive. It was known that the resistivity of the metal varied with the amount of light falling upon it, but that selenium itself could actually turn light into electricity was an amazing discovery. It was left to an American inventor called Charles Fritts to make something practical of this little-understood phenomenon. He did so in 1883 by coating selenium with a very thin plating of gold, so thin that it was transparent to light. Fritts found that, like any other electric battery, the cells could be arranged in an array to increase the output of electricity.

Because he did not really understand how his solar cell worked, Charles Fritts contacted an expert in the field, Werner von Siemens. After replicating Fritts' work and finding that the cell really did produce electricity simply from exposure to light, von Siemens was very optimistic about the future of photovoltaic arrays. He wrote a paper on the photovoltaic effect, concluding it with the following words;

> In conclusion, I would say that however great the scientific importance of this discovery may be, its practical value will be no less obvious when we reflect that the supply of solar energy is both without limit and without cost, and that it will continue to pour down upon us for countless ages after all the coal deposits of the earth have been exhausted and forgotten.

The message was clear: fossil fuels were a thing of the past and solar power was the obvious choice for the future, when it came to producing electricity.

To the average person living in the developed world at that time, Charles Fritts' invention and Werner von Siemens' advice must both have seemed preposterous. Selenium and gold were rare and expensive, whereas oil and coal were plentiful and cheap. Why on earth would one fool around with fancy devices in their experimental stage, when by burning coal and using it to boil water and spin a steam turbine, one could produce all the electricity that was needed at once? This is an attitude which is, unfortunately, still going strong to this day. Surprisingly though, there were at that time already people who were beginning to worry about that most modern of concerns, climate change, and hoping to avert it by turning away from fossil fuels and embracing the emerging technology of renewable energy.

Most of the calculations relating to man-made climate change, the figures which we use to try and predict what will be happening to the earth's weather in 50 or 100 years, if we continue to burn fossil fuels, were carried out not, as one might guess, in the 1990s but rather in the 1890s. They were the work of one dedicated man.

Most people spend Christmas Eve with their family or friends, preparing for the what is, to many people, a very special day. Swedish polymath Svante

Arrhenius, however, did not feel at all festive on 24 December 1894. His marriage had collapsed and it was cold and dark in Sweden at that time of year. Instead of visiting the neighbours, attending a carol service or indeed seeing anybody at all, Arrhenius chose that day to begin work on the subject of carbon dioxide levels in the atmosphere and how they might affect global temperatures. He was to spend a year on this endeavour, which culminated in his presenting a paper to the Stockholm Physical Society entitled, *On the Influence of Carbonic Acid in the Air Upon the Temperature of the Ground.* In other words, what effect the amount of CO_2 in the atmosphere has on how warm it is.

Initially, Arrhenius was curious to know what might have triggered the Ice Ages, but he soon realized that the figures which he was working on might have relevance to the future, as well as the past. After concluding that a reduction of CO_2 levels would cool the earth overall by 4 or 5 degrees Celsius and would in itself be sufficient to cause the whole of Europe to be covered in ice, he wondered what would happen if future CO_2 levels were to rise significantly. What if they were to double, say? His calculations, which were all done by hand and took him the better part of a year, suggested that if the amount of carbon dioxide in the atmosphere were to increase that much, then the temperature of the planet would go up by between 5 and 6 degrees Celsius. These figures match exactly those which the Intergovernmental Panel on Climate Change came up with, after feeding all their data into a super-computer.

It is perhaps not entirely a coincidence that Svante Arrhenius went on to become a determined advocate of renewable energy. He became obsessed with the idea that fossil fuels, oil in particular, were likely to run out and that the only answer for humanity's survival lay in solar energy and hydroelectric power. His influence was considerable; he was no crank, but rather a Nobel Prize-winning scientist. After sitting on a government commission looking into the providing of electricity to Sweden, he was instrumental in ensuring that the country became the first to exploit hydroelectric power on an industrial scale.

Chapter 8

The Resistible Rise of the
Internal Combustion Engine

S teampunk novels contain any number of strange devices which in the real world turned out to be irrelevant dead-ends; semaphore telegraphs, mechanical computers and steam planes, to name but a few. These curiosities frequently feature prominently in steampunk stories and are often an integral part of the plot. Mechanical computers, for instance, drive the entire narrative of *The Difference Engine*. In *Pavane*, the semaphore telegraphs are mentioned in every chapter but one. The one invention which seldom or never appears in steampunk fiction is the internal combustion engine. Even when, as in Harrison's *A Transatlantic Tunnel, Hurrah!*, there are such advanced modes of transport as railway trains powered by nuclear energy, the ordinary petrol-driven motor car is wholly absent. Sometimes, as in *Pavane*, the absence of cars and other petrol-driven vehicles is explained, in that case by a prohibition of the Catholic Church, designed to impede technological progress. In other books though, the lack of internal combustion engines is merely noted and taken as a given.

The alternatives to the role assumed in our own world by petrol or diesel-powered lorries, buses, cars and taxis varies greatly. In *Bring the Jubilee*, 'minibiles' powered by steam engines are in use, rather than the automobiles with which we are familiar. Steam cars are also to be found in *The Difference Engine. A Transatlantic Tunnel, Hurrah!* has, in addition to steam cars, Hansom cabs which are drawn by battery-powered electric motors, while in the *Mortal Engines* books of Philip Reeves, a wholly novel mode of travel is found. In the moving city of London described in *Mortal Engines*, hot air balloons are found to be the most convenient way of getting from one part of the city to another.

Motor cars, buses and lorries with petrol engines are such a part of everyday modern life, that removing them creates at once a disconcerting sense of strangeness in a fictional world set in an alternative present. In *The Difference Engine*, it seems very right and proper that although there are racing cars, these should be powered by steam. When we read of an alternative version of the 1950s, 1970s or the present-day world which lacks cars with internal combustion engines though, their absence is very noticeable. The very idea that we would, in the twentieth or twenty-first century, be reliant upon steam for transport tells us at once that we are in a very different universe from our own. Part of this sense of dislocation is caused by our own faulty perception of the nature of history.

In retrospect, any version of history other than that which actually occurred, has an air of fantasy about it. We look back on the way that the world developed and see it as being somehow inevitable and the only logical and sensible course that events could have followed. It is enjoyable to be entertained by accounts of unfamiliar versions of the past, where steam cars became the dominant form of personal transport and the petrol engine was viewed as a technological dead-end, but that of course is merely the stuff of extravagant and impossible daydreams. It is obvious that once we no longer relied upon horse-drawn vehicles, then animal power would have to be replaced and the only logical replacement must surely be vehicles powered by oil. The truth is that the rise of the internal combustion engine was not at all inevitable and its ascendency the result of a concatenation of chances and random happenings, rather than the force of history at work.

As the end of the Victorian era drew near, it was plain that horse-drawn transport had had its day in Europe and America. Horse and carts were slow and clogged the streets of the rapidly-expanding cities until traffic was slowed to a virtual standstill in big cities such as London. As early as 1855, the traffic jams in London made travel by any kind of vehicle only possible at a snail's pace. That month, the Parliamentary Select Committee on Metropolitan Communications was sitting, with the aim of trying to find a way of easing congestion in London. Joseph Paxton, who had designed the Crystal Palace and was by then a Liberal MP, was called to give evidence. He told the committee that so slow was travel by road in the capital that it was quicker to travel by railway from Brighton to London Bridge station than it was to travel from London Bridge to Paddington, another London railway station.

Not only were horses a slow way of moving traffic about, they were also vulnerable to disease. Just how vulnerable was demonstrated in the autumn of 1872, when an epidemic of equine influenza began in Canada and spread rapidly to the United States. Although the mortality rate was less than 10 per cent, animals afflicted with this virulent strain of influenza were unable to work for a fortnight or so. Since 95 per cent of horses were affected, this meant that fire engines were unable to answer calls, deliveries of food were interrupted and public transport in cities ground to a halt. There were no longer any trams, buses or taxis, all of which were drawn by horses. What became known as the Great Epizootic was a wake-up call and showed that reliance upon horsepower in industrialized countries was unwise.

After the epidemic died away, a search began for new ways of moving people around cities; specifically, a means of public transport which did away with the need for horses. Steam trams were tried, as well as carriages pulled by cables. Experiments were also conducted with electricity sent along rails or overhead wires. It was some time before a successful method was devised, but by the end of the century, electric trams were becoming common in both America and Europe. This was all well and good for moving passengers along main routes through cities, but it still left delivery vans, fire engines, taxis and private carriages reliant upon horses. The search was on for a reliable and effective way of propelling vehicles of any kind by means of a self-contained source of power.

There were, as the Victorians approached the end of the nineteenth century, three contenders for the best and most efficient way of driving vehicles in the absence of horses or oxen; these were steam, electricity and the newly-invented internal combustion engine. Of these, the internal combustion engine appeared the least likely candidate to be generally adopted. Steam engines had been around for well over a century and many people already had experience of travelling by steam, either in trains or boats. The technology was highly developed and familiar to everybody. Electricity was a relative newcomer, but electric trams were becoming more common. Many people saw electricity as the motive power of the future. 'Electric' meant up-to-date and modern; the word had the same connotations which 'atomic' would acquire in the 1950s. If steam would not fit the bill for the replacement of horsepower, then electricity would surely provide the answer. Trailing behind as the least likely contender for the power source to revolutionize transport was the internal combustion engine. To see why the widespread adoption of petrol engines seemed so improbable in the late nineteenth century, we need to look in some detail at steam, electricity and petrol as they were being used at that time to move vehicles. We will start with internal combustion engines which burned petrol.

Before going any further, it might be as well to explain why we talk of 'internal combustion engines' and how they differ from any other kind of engine. Combustion, of course, means no more than 'burning'. In a steam engine, the generation of heat by burning coal, wood or some other material takes place outside the engine; which runs on water. In the internal combustion engine, the fuel is burned inside the engine and itself provides the energy, without the intermediary stage of heating water. Petrol, diesel, alcohol or gas is compressed and then catches fire inside a cylinder.

The idea of the internal combustion engine had been around for some while before the beginning of the nineteenth century. The only problem was, nobody had been able to find a way of actually *making* one. In the late seventeenth century, Dutch polymath Christiaan Huygens, who invented the pendulum clock, drew up plans for an internal combustion engine running not on petrol but gunpowder. This was quite logical, if one wanted at that time a powerful source of energy. The gunpowder being detonated, the force of the explosion acts to push up a piston, so providing the motive force for a machine. For the next 150 years, the idea of an engine powered by gunpowder circulated, with nobody able to overcome the technical difficulties of constructing such a device.

In 1800 a London clockmaker called George Medhurst was granted a patent, No. 2431 that year, for a 'New, Improved Method of Driving Carriages of All Kinds, without the Use of Horses'. Most of Medhurst's patent was taken up with a system for using windmills to create reservoirs of compressed air, which carriages would draw on, much as cars today visit a filling station for petrol. The compressed air would then drive a motor and power the vehicle. On the last page of the technical drawings though, almost as an afterthought, was a detailed plan for an internal combustion engine, which would use gunpowder as a fuel. There is no record of such an engine ever being built and it is doubtful if it would have worked.

Soon after George Medhurst toyed with the idea of driving a carriage by gunpowder, George Cayley, the pioneer of flight at whose work we looked in Chapter 4, also turned his attention to a gunpowder engine. He was looking for a way to propel an aeroplane and had found that steam engines were too cumbersome and heavy for the purpose. He needed a light, powerful engine which could turn a propeller fast enough to keep an aeroplane aloft. In a letter, Cayley claimed to have put together a small engine which ran on gunpowder, but admitted that it did not work very well. He later, according to his records, used one in a model aeroplane.

Another solid substance chosen to power early internal combustion engines was coal and it proved a great deal more successful than gunpowder. The Niepce brothers, who went on to invent photography, built the world's first working internal combustion engine. This they named the Pyreolophore and in 1807 they demonstrated that it was capable of propelling a boat upstream, that is to say against the current, on a river. The fuel was powdered coal, mixed with both resin and dried spores from moss. Twenty years later, those master engineers of the Industrial Revolution, Marc Brunel and his son Isambard Kingdom Brunel, also tried to construct a new kind of engine, this one running on liquid carbon dioxide. What they called their 'gaz engine' was a flop and after ten years, they gave up on the project.

The first successful internal combustion engine was made by the Belgian inventor Etienne Lenoir in 1859. He converted a steam engine so that instead of water, it ran on a mixture of coal gas and air. These vapours exploded in the cylinder and so drove a piston down to turn a crankshaft. The Lenoir engine was a great success and many were sold throughout Europe. Three years later, Nikolaus Otto in Germany devised the four-stroke engine which most modern cars incorporate, with four cylinders which work on a cycle of compression, ignition, power and exhaust. Otto's engine, like that of Lenoir, ran on coal gas. These engines relied upon a spark setting fire to the explosive mixture when it was fully compressed. A few decades later Rudolf Diesel, another German, discovered that some distillates of oil would self-ignite, if sufficiently compressed. This resulted in the development of the type of engine which today bears his name. Before this, Diesel experimented with using coal dust as a fuel.

One thing which readers will probably have noticed is the variety of curious fuels upon which these early engines ran. Gunpowder, coal dust, carbon dioxide, coal gas; the one thing missing is petrol. There is a very good reason for this. Until about 1859, oil was obtained only by finding a suitable spot where it was seeping naturally from the ground and could then be scooped up in buckets. The first commercial drilling for oil and subsequent realization that pumping it up out of the ground in large quantities was a practical proposition, is usually acknowledged to be the oil well set up in Pennsylvania by Edwin L. Drake, between 1857 and 1859. Once it was realized that this was an effective way of extracting oil from the ground, others quickly emulated Drake's methods and the era of plentiful and cheap oil had arrived.

It was the sudden flood of oil, which it was soon found could be readily distilled into petrol, which caused those developing internal combustion engines to see these as the motive power of the future for road transport. There were, however, many problems to be overcome. The eventual triumph of the petrol-driven internal combustion engine was far from inevitable as the nineteenth century drew to a close, which may be seen by glancing at both the land speed records for self-powered wheeled vehicles for the opening years of the twentieth century and also seeing what different kinds of motor cars were being produced and bought.

We tend to think of the Edwardian Age, the years between the death of Queen Victoria in 1901 and the outbreak of the First World War in 1914, as being the time when motoring took off. We are all of us familiar with the image of the early motorist, wearing gauntlets, goggles and a peaked cap, driving along country roads and causing alarm to quiet rural districts. These were the days when the internal combustion engine began to come into its own and ushered in a familiar part of the modern world. Except of course, that this is not at all what happened and it was touch and go at that time whether the petrol-driven motor car began a triumphant rise or instead fell by the wayside as just one more footnote in history.

Readers of *The Difference Engine* are entertained to read of a racetrack where speeding cars compete for the trophy. One vehicle in particular is notable for its radical new design; it is streamlined, to reduce the friction of the air as it speeds along the track. The twist is of course that these racing cars are the steam-powered successors to Gurney's coach, which we looked at in Chapter 1. (In another little burst of inventiveness, the word 'streamlined' is altered to 'line-streamed'.) This makes the whole affair pleasingly fantastic and bizarre. Imagine a streamlined car powered by a steam engine! History shows though that there is nothing in the slightest degree preposterous about such a notion.

The land speed record is the highest speed attained by a person using a vehicle on land. Railway trains do not count. In 1898, the land speed record was 39.24mph. The following year, it had crept up to 65.79mph and by 1902 it had jumped to 75.06mph. The interesting point is that the first two of these records were achieved by electric cars and the one in 1902 by a steam car. At a time when most of us imagine the internal combustion engine was forging unstoppably ahead and overtaking the old horse-drawn transport, electricity and steam are actually at the cutting edge of developments in self-propelled vehicles, rather than petrol engines.

Two targets in particular entranced early pioneers of speeding cars. One was to develop a car which would travel faster than any train and the other was to break the 200km/h barrier. In 1906, a car reached the astounding speed of 127.66mph; a little over 200km/h. This was far faster than any railway train of the time; it would be over thirty years before a train exceeded a speed of 200km/h. The car which smashed through these two barriers on 26 January 1906 was powered by steam.

Still, it may be argued, perhaps specially modified or adapted steam cars, or those running from batteries, might be able to perform in this way, but surely on the roads, most cars were using internal combustion engines? Not a bit of it. As the twentieth century began, almost 80 per cent of the cars on American roads were *not* fitted with petrol engines. The figures are quite unbelievable; 40 per cent of cars at that time used steam engines, 38 per cent were powered by electricity and only 22 per cent relied upon internal combustion engines. In the race to replace horse power, the petrol-driven internal combustion engine was lagging far behind. The most popular and best-selling car in America in 1900 was the Columbia Electric Runabout, which was also, incidentally, the first American car to reach sales of over 1,000. Even those whose business was to predict future trends were wholly unable at that time to foresee the inexorable rise of the internal combustion engine. This was the case in Britain, just as it was in America and the rest of the developed world. With electric cars, taxis and buses, to say nothing of steam cars which were breaking the land speed record, how could anybody know that petrol-driven cars would one day conquer the world?

The Royal Commission on London Traffic had met in 1905 to consider the best way of easing congestion on the streets of the capital. Some witnesses advocated more buses fitted with internal combustion engines as a way of reducing traffic jams. The report produced by the commission was scathing about the idea. It read, in part, 'We cannot recommend the postponement of tramway extension in London on the ground of any visible prospect of the supersession of tramways by motor omnibuses.' Which, in the archaic language of the time, was as much as to say; don't bother about these new-fangled motor buses, they won't last. In 1909 Aubrey Llewlyn Fell, Chief Officer of the London County Council Tramways, was even more explicitly dismissive of the idea that buses with internal combustion engines would ever be useful. He said, 'Twenty years hence, motor buses will be exhibited as curios in museums.' Why were petrol engines so unpopular in the United States in the years leading up to the First World War and why did those planning for the future of transport in London regard buses powered by such engines as being of a negligible importance?

To understand why the internal combustion engine nearly failed to become the favoured method for propelling vehicles of all kinds, we have to go back to the 1880s and 1890s to weigh up the relative advantages and disadvantages of the different kinds of cars on the roads. When we do so, we might see exactly why petrol engines were seen as being so unpromising at that time.

One of the chief problems to begin with, with early cars fitted with internal combustion engines, was that they were forever breaking down. Unlike steam engines, which had been around for over a century and undergone continuous development and refinement over the whole course of that time, engines running on petrol were novelties. The technology was new and the teething problems were many, varied and significant. It must be borne in mind that although we talk of internal 'combustion', we really mean internal 'explosion'. When compressed petrol or diesel vapour ignites; there is a sudden bang, rather than the steady flame which is used to run a steam engine. Because modern vehicles are fitted with complex and efficient mufflers, silencers and sound-suppressers, we tend to forget that they are pushed forward by a series of violent explosions, producing loud bangs. Anybody who has ever heard an ordinary car engine running when the silencer has been removed will be aware that it sounds like a machine gun; a deafening and constant crash of explosions. This racket contrasted unfavourably with the almost noiseless operation of small steam engines. Early internal combustion engines were both exceedingly noisy and also very smoky and smelly.

It was Hiram Maxim, whom we saw in an earlier chapter launching his gigantic steam-powered aeroplane in Kent, who eradicated one of the main disadvantages of the petrol engine, that of the terrible noise which they made. He did this by inventing what we call today the muffler or silencer. All the exhaust gases, and consequently much of the noise, of the internal combustion engine are directed through a long pipe which runs the entire length of the vehicle. Along this tube are baffles and padding which tend to absorb or deflect sound. The familiar purr of a well-tuned car engine, with which we are today familiar, is produced by some version of Maxim's muffler. Without it, the average car engine would sound like a heavy machine gun being fired near at hand. Along with the electric starter motor, the muffler made petrol engines popular and brought about the decline of the steam and electric car.

The unreliability of the early petrol driven cars was legendary. They broke down so often that every driver needed to have the skills of a mechanic; there were of course few garages or service stations. Cartoons in *Punch* magazine regularly showed men, and occasionally women, with broken-down cars by the side of the road, tinkering with the engine to try and get it to work again. One cartoon consisted of two frames. In the first, an old woman driving a pony and trap is being overtaken by a motor car full of fashionably-dressed people, who cast supercilious glances at her as they pass. In the next, she is smiling slyly as she passes the motor car, which has now broken down, while the driver and passengers work on the engine.

Perhaps the best way to see how ordinary people viewed the early cars powered by internal combustion engines is to look at the themes of postcards on sale at that time in Britain. Postcards were, in the early days of motoring, the most popular way for relatives and friends to keep in contact in Britain. In 1908 alone, 860 million picture postcards were sent in the United Kingdom, which meant that some two and a half million were being posted every day of the week. Because this method of communication was so tremendously common, the images on postcards can give us a very good idea as to what ordinary people were interested in and how they felt about various matters. The cartoon-like pictures which we see featuring cars are exceedingly revealing.

The first thing which strikes us when looking at postcards showing motor cars is that there were an awful lot of them. The fact that images of cars driven by internal combustion engines appear so frequently indicated the level of public interest in the subject. Just as crazes such as roller skates or Diablo tell us that these were things which were very much in the consciousness of the age, so too do all those humorous pictures of cars tell us that ordinary people had them in their minds.

The general impression one gains from picture postcards of the early days of motoring is that motor cars with petrol engines were thought to be unreliable and dangerous; accident-prone, liable to break down, capable of blowing up, and noisy and smelly into the bargain. Many cards show cars having to be pushed or towed home, after they have stopped working. One postcard depicts the front wheels of a car coming away from the body of the vehicle, as it travels down a hill. In another, a man is seen walking down the road, while in the distance a car is exploding; hurling the driver into the air. The caption reads; 'DEAF OLD GENT: THERE, THAT'S THE FIRST TIME I'VE HEARD A NIGHTINGALE THIS YEAR'. Another has a policeman talking to a man with a stick, while a motor car disappears in the distance. The text reads as follows; 'POLICEMAN, INADVERTENTLY ACCOSTING BLIND MAN - "HAVE YOU SEEN A MOTOR PASS THIS WAY?" B.M. "NO, BUT I SMELT ONE."'

Petrol cars were regarded as noisy, smelly, dangerous and unreliable novelties; very different from the familiar steam engine. This point of view was neatly summed up in the early days of the internal combustion engine in a magazine which, although started to promote motoring, recognized that most people disliked the idea of petrol-driven cars.

Automotive Industries is the world's longest continuously running trade magazine, having been founded in 1895 with the title *Horseless Age*. The year after it started publication, *Horseless Age* freely admitted that the internal combustion engine faced an uphill struggle for acceptance. In an editorial, the following words appeared;

> The vast majority of people would prefer a smooth-running, reliable steam engine to the evil-smelling, dangerous, wasteful and at best uncertain and unreliable internal combustion engine.

Unlike the steam engines of the time, all petrol engines at the beginning of the twentieth century were more or less experimental, meaning that drawbacks became apparent after a few months or years. With no radiators or fan belts, such engines were prone to overheating. On cold mornings, it was sometimes necessary to use a hot iron to warm up the cylinders before the engine would start. One particular hazard was that to start the engine, somebody had to place a 'starting handle' through a hole at the front and turn it, to get the cylinders moving up and down. The crank frequently took on a life of its own and unexpectedly begun moving violently in the opposite direction, resulting in broken collar bones and wrists or bruises to the face.

Another factor which influenced prospective buyers of cars and served to discourage some of those who might otherwise have been inclined to choose one with an internal combustion engine was the idea of travelling around with a tank full of highly-inflammable liquid. This seemed a risky business; after all the petrol could even, as the operation of any engine demonstrated, explode under the right circumstances. All in all, investing in one of the new cars of this type was a chancy and uncertain business. How different was the case for those who chose a vehicle fitted with a steam engine.

Steam cars were clean, safe, efficient and silent. Nobody ended up with a broken wrist through turning a starting handle at the front of the engine, as all too often happened with petrol-powered internal combustion engines. True, there was a slight delay at the beginning of a journey, as one waited for the engine to fire up, but this minor inconvenience was more than compensated for once the journey was under way. With early internal combustion engines, it was necessary to slow down the car in order to change gears. This was not the case with steam cars; there, one merely pressed down on the accelerator and the car's speed increased with no fuss or bother at all; there was no gearbox. This instant acceleration was one of the selling points of steam cars and had they not been eclipsed by the cars running on cheap oil, then such positive features as swift acceleration and a lack of gears would certainly have become increasingly attractive.

To see the potential which steam cars had, we need only look at the models which were still being produced a few years later, after the end of the First World War. We saw how a steam car held the world land speed record at the start of the twentieth century. The same company which made this vehicle, Stanley, produced a range of ordinary cars known as the Stanley Steamers. It was another American company though which really took the steam car to the limits of its technical potential and it is, at first sight, something of a mystery why sales of their cars did not overtake those running on the increasingly common internal combustions engine. Perhaps it was because as the twentieth century progressed, the word 'steam' was already acquiring the negative connotations which it has today; that is to say as being synonymous with 'backward' and 'old-fashioned'. There was certainly nothing backwards or old-fashioned about the Doble company's cars which were made between 1909 and 1931 and showed clearly that the steam cars so popular in steampunk stories were a perfectly practical proposition.

The four Doble brothers, Abner, William, John and Warren, came from a family with an interest in engineering; their father had become wealthy from patenting a new kind of water turbine for hydroelectric power plants. At a time when cars driven by internal combustion engines had begun their seemingly unstoppable rise to world domination, the Doble brothers began to devote their energies to producing new and improved steam cars.

In the years leading up the start of the First World War in 1914, the Dobles produced several experimental cars which were at least as good as, and in some respects superior to, the petrol-driven cars which were on the market, including Henry Ford's famous Model T. The first of these improvements was a steam condenser. Typical steam cars at that time had to be filled with water every 20 or 30 miles. By condensing the steam which had been used and returning it to the tank, the Dobles' car could travel for 1,500 miles without needed to take on any water. There was no clutch or transmission, which made it easier to operate and, most interestingly, its acceleration was better than the majority of petrol cars of the time. Ford's Model T took 40 seconds to reach 50mph from a standing start, the car made by the Dobles could go from 0–60mph in just 15 seconds.

Although most manufacturers had decided by the time of the First World War that the future of cars lay in petrol, rather than steam, there was considerable interest in the work of the Dobles. At the 1917 New York Motor Show, there was great enthusiasm for the Dobles' latest model. This had a key ignition, which meant that there was no fiddling around lighting the burners manually. The water was heated by kerosene, which was cheap and easily obtained. Apart from the technical specifications, which rivalled those of any car at the show powered by an internal combustion engine, the best bit about the Dobles' car was that there were only four controls. These were the steering wheel, brake, throttle and a cut-out, so that the car could be put into reverse.

In the 1920s, the Dobles produced their most sophisticated models; cars which could start from cold in 30 seconds and comfortably reach a speed of 90mph. Enthusiasts working today with cars made by the Dobles at that time, have managed to get them to travel at 120mph and all in almost complete silence.

Readers are perhaps wondering at this point why these wonderful vehicles did not overtake the internal combustion engine in sales, if they were so efficient. The answer is simple. Because the Dobles were turning out their cars themselves, without the mass-production, conveyor-belt system adopted by bigger companies such as Ford, the cars were much more expensive. Most people would sooner have a cheap car which was not perfect, rather than a beautifully-crafted one which was extremely expensive. After all, the mass market for cars was really as runabouts for families. The ability to go from 0–75mph in 10 seconds or to cruise at 90mph were not really selling points for the average buyer. True, they were very quiet, but this has never particularly been something about which the average buyer of a motor car is concerned. It is perhaps significant that the only time that the silence of a car engine has been used as a selling-point has been with that epitome of motoring luxury; the Rolls-Royce. Readers will probably be familiar with the 1957 advertising slogan for the company, 'At 60 miles an hour the loudest sound in the new Rolls-Royce comes from the electric clock.' This is not the sort of thing which would sell more Fords, because those wanting a cheap family car didn't really care how much of a racket it made. The Dobles' cars were strictly for the very limited luxury market and this was their downfall.

The Dobles eventually went out of business in 1931 and their plant and materials were acquired by another set of brothers, George and William Besler. The Beslers had the idea of fitting one of the Dobles' steam engines into an aeroplane, as was mentioned in an earlier chapter. Just as with steam cars, there were certain advantages to steam planes, such as the ability to change the direction of the propellers almost instantaneously, so making the plane able to land in a very short distance. Steam planes were, like cars, almost silent in operation, which might have had some military potential; they could have become the first 'stealth' aircraft.

It would not have taken more than a slight change in history to have made the steam car the people's choice in the years after the end of the First World War, but men like Henry Ford had realized that cheapness was more important than silence or even reliability and his production-line methods certainly brought down the cost of a family car. It was this which spelled the end for the steam car as a popular means of travel, rather than any technical considerations.

There was one area in which steam-powered road vehicles held their own though, as late as the 1930s. We saw in Chapter 1 that the steam bus services which were running before Queen Victoria began her reign had been driven out of business by a combination of sabotage and artificially-inflated road tolls. There was nothing inherently unsound about the steam buses which plied the streets of London, Gloucester and Edinburgh; it was merely that other commercial interests managed to squeeze them out of existence. Much the same thing happened to the steam lorries which flourished in Victorian Britain and even lingered on until well into the twentieth century.

In *Pavane*, set in a late twentieth-century England where electricity and internal combustion are all but unknown and there are no railways, the movement of freight is undertaken by steam wagons. Like the network of semaphore towers upon which long-distance communication depends, the steam locomotives which haul trains of trailers along the roads, from one city to another, are a recurring theme in the novel. The names of the various manufacturers such as Robey, Fowler and Foden are mentioned and this, combined with long descriptions of the technical aspects of steam traction engines, enables Keith Roberts to build up a fantastically-detailed world in which the petrol engine never quite got going. Just as with the optical telegraph that forms another of *Pavane*'s recurring themes, the road steamers required less imagination on Roberts' part than they did a sound grasp of industrial history.

Until the abolition of what is often known as the 'Red Flag Act' in 1896, any self-powered vehicle on the roads of Britain was obliged by law to have a man walking ahead of it with a red flag, to warn people of the approach of that most alarming of phenomena, the carriage not drawn by a horse. It need hardly be remarked that this, combined with a speed limit of 4mph, greatly hindered the development of any sort of road vehicle, whether powered by steam, electricity or an internal combustion engine. In November of that year, the speed limit was raised to a dizzying 14mph and the need for a 'footman' was done away with. It was certainly good news for the handful of British motor car enthusiasts, who celebrated by organizing the first London to Brighton rally, but it was also an exciting opportunity for road haulage.

A number of companies had been experimenting with steam lorries, or steam wagons as they are more generally known. Nobody in his senses at that time would have relied upon the new and largely untested petrol engine to transport goods around the country, but steam engines were familiar and not liable to break down at every touch and turn. The 1890s saw an explosion of interest in moving freight about by steam wagon. The companies involved in making these new lorries were the ones mentioned repeatedly in *Pavane*, Foden and Fowler, for example.

There had for years been trailers towed by traction engines, but now there was an opening for what we today call lorries, with a cab at the front and space for a load at the back. For almost 20 years, these steam wagons became a major feature of road transport. The First World War resulted in manufacturers focusing on tractors and wagons which would be of use for pulling artillery and so on, but after 1918 the steam wagons began to compete seriously with petrol lorries. The 1920s were a golden age for steam haulage on the roads of Britain, but it was not to last. Two factors brought about its demise by the start of the Second World War in 1939.

There was a perception that by not putting a hefty tax on road haulage, the government was in a sense subsiding those competing with the railway. Of course, during the Great Depression, a new way of raising money for the government was welcome in any case. In 1933, a tax was imposed on road hauliers and it was calculated according to axle weight; the unladen weight of the vehicle. This Axle Weight Tax hit steam wagons more heavily than it did those with internal combustion engines because of course steam engines are much heavier than those which use petrol, due in part to the sheer weight of water which must be carried, in addition to the fuel needed to heat it up. The tax came as something of a shock to hauliers operating fleets of steam wagons. The knockout blow came the following year, when the road fund tax on road locomotives was raised to £100, while at the same time there was a reduction in taxes on fuel oils, which were imported from abroad. This made petrol engines cheaper at a stroke.

By 1939, steam wagons had all but vanished from the roads of Britain and the internal combustion engine was triumphant. This triumph was due at least as much to the meddling of the state as it was to any technical superiority of petrol and diesel over steam. One great advantage of steam for the motorist has already been mentioned. It is that no gearbox is needed. One simply increases the rate of the engine and greater speed results automatically. Before we turn to look at what was happening with electrically-driven cars, buses and taxis at the end of the nineteenth century, we must look at one mode of road travel which one would, at first sight, assume could only have been the creation of a steampunk author; the steam motorcycle.

It is commonly claimed that the world's first motorcycle was that put together by Gottlieb Daimler in Germany and patented on 29 August 1885. Photographs of this motorcycle, which was powered by a four-stroke engine, are to be found in books listing the first appearance of various inventions. The truth is somewhat different. If we define a motorcycle as a self-propelled, engine-powered, two-wheeled vehicle, then one exists to this day in a museum which predates Daimler's bike by over 15 years.

The Smithsonian Institution's National Museum of American History describe one of their exhibits as 'the jewel of the Smithsonian's collection'. It is a steam velocipede, invented by Sylvester Roper in the 1860s. Nobody knows the precise date that Roper perfected the machine, but by 1869 it was being demonstrated at circuses and fairs in New England. Although this, the world's first motorcycle, looks a little crude and ungainly, it has some very modern features. Twisting the grip of the right handlebar, for instance, opens the throttle and increases the speed, just as with modern machines. The left-hand grip of the handlebars was twisted to apply the brake.

Sylvester Roper went on to refine his steam motorcycle and produce more efficient versions of it, incorporating the frame of a safety bicycle. In 1896, at the age of 72, Roper offered to put his latest machine through its paces at the Charles River Velodrome in Boston, a custom-built concrete track for bicycle racing. After showing how easily he could outpace professional cyclists with his own bike, being timed at around 30mph, Roper was invited to test his motorcycle's full speed. The track was cleared and the old man began racing around the track at a little over 40mph. Then, his face was seen to go very white and he turned off the power to his bike, slowed down and died of a heart attack. With him died the idea of steam motorcycles, for nobody showed any interest in developing a practical model following Roper's death.

We regard electric cars today as the epitome of modernity. The British government is fantastically keen on the idea of doing away with petrol engines and plans are afoot to provide public charging points in cities and new residential developments to encourage the use of such vehicles. The electric car is, we are led to believe, the future of transport. It is confidently asserted in some quarters that in a few years the ordinary motor car will have become a museum-piece. There is, however, something eerily familiar about the idea of cities where the loudest sound is the hum of an electric motor and there are no exhaust fumes from buses, lorries and cars to contend with. It seems familiar to some of us, because we are aware that this was how things began to be at the end of the Victorian era.

With the abolition of what we know today as the 'Red Flag Act', the way was open for self-propelled vehicles to take freely to the roads of Britain. Steam cars and wagons were one manifestation of this new freedom; another was the proliferation of electric cars. Sometimes a word comes, in the minds of ordinary people, to represent all that is modern or futuristic. In the 1950s and 1960s 'atomic' or 'nuclear' were the words to conjure with and anything described as 'atomic' was automatically understood to be state-of-the-art or cutting edge. In the closing years of Queen Victoria's reign, much the same connotations were attached to the word 'electric'. Streets in London were named 'Electric Avenue' or 'Electric Parade' to show how up-to-date and modern they were. When Robida wished to write a romance of the future, set a century hence, he chose to call it *The Electric Life*, because that way his readers would know at once that this was likely to be a futuristic fantasy.

All of which meant that electric cars sounded like the future of transport. Apart from the semantics of the word, the cars themselves were quiet and reliable, unlike the petrol cars with which they were competing. Nor did electric cars have that air of vulgarity about them that the internal combustion engine brought to mind. Those driving an electric car did not find the need for the goggles, gauntlets, scarves and caps which were associated with the motor car. They were altogether more genteel and sophisticated. This too worked to the advantage of those selling them. Quite a few members of royalty and the aristocracy, both British and European, drove electric cars. Such people would not have been seen dead racing around in a noisy, smelly petrol-driven car. In 1901, the year that Queen Victoria died, the City and Suburban Electric Carriage Company published the names of some of those who had purchased their cars; Queen Alexandra, Edward VII's consort, the Dowager Empress of Russia, the Marchioness of Ripon, the Princess de Polignac, Prince Demidoff, the Duke of Sutherland, the Earl of Derby; the list of noble names was a long one. Selling one of their cars to the Queen of Britain was a smart move on the part of the City and Suburban Electric Carriage Company, for it enabled them to put the royal coat of arms on their publicity material, beneath which was the legend; 'By Special Appointment to Her Majesty Queen Alexandra'.

The electric cars which were on the market at this time had specifications which might do very well for a little smart-car for use in the city today. The top speed of 25mph would not prove too much of a handicap in London traffic and the range of 80 miles, which some models promised, compares favourably with modern electric cars. The batteries on which the cars depended could be recharged in around three hours.

Looking at the exhibitors in the Automobile Exhibition held at Crystal Palace between 30 January and 7 February 1903 is a disconcerting experience. Some of the names are recognizable, we see Daimler and Wolseley, but also the names of companies which produced only steam-powered vehicles or electric cars. For instance, we find that the City and Suburban Electric Vehicle Company had a stand there, as did Brush Electric Engines, Electromobile and Thorneycroft Steam Wagons. Any visitor to the Automobile Exhibition that year would not have been able to guess which sort of cars would be the big thing of the twentieth century. Steam and electricity would have seemed just as likely as internal combustion engines.

Electric cars had already been around for decades when the Automobile Exhibition was held in 1903. Sales were slow at first for several reasons, not least of which was that domestic electricity supplies were not common. If your home was lit by gas, then you would not be in a position to plug in your car and recharge its batteries. This problem too had been thought of and it led in 1901 to the first example of a very modern kind of building; the multi-storey car park.

The City & Suburban Electric Carriage Co had a seven-storey car park built at 6 Denmark Street in London's Soho district. Motor cars were at that time so rare that it had never before occurred to anybody that it might be necessary to have special buildings of this sort to park them in. This multi-storey carpark was used to keep the cars belonging to people who did not have electricity in their homes. They could be charged there overnight or left for weeks or more, if needed. For a small additional charge, the company would deliver the fully-charged vehicles to the homes of their owners.

The rarity of mains electricity was certainly a limiting factor in the popularity of electric cars. Since they ran on rechargeable batteries, these would need to be changed regularly, when they ran down, for fully-charged ones. Without sockets in one's home, it was a little difficult to run such a vehicle. In 1910, only 2 per cent of British homes had electricity and it was not until after the end of the First World War that this situation began to change to any great extent. One company involved with electric cars solved the problem by cutting the Gordian Knot and creating their own electricity supply.

At the South Kensington Motor Show in 1896, an exhibitor called Walter Bersey unveiled what he hoped would be the wonder of the age; the electric taxi cab. Bersey hoped to cash in on the abolition of what has become known as the 'Red Flag Act', which was due to come into force that November. With the speed limit for self-propelled vehicles raised from 4mph to 14mph, he felt that he could make a killing with a fleet of electrically-powered cabs, which would render the horse-drawn Hansom cab obsolete overnight. There was great interest in the idea and the beginning of the service was awaited with interest. The old speed limit and requirement for the 'footman' with the red flag ended on 13 November 1896 and the following year, the taxis duly appeared on the streets of London. There had been some teething troubles, as is so often the case with new inventions which are revealed to the public too soon, and it was not until 19 August 1897 that the first of the new taxi cabs began plying for hire.

The Metropolitan Police, like everybody else, were unsure what to make of the new cabs. They laid down four rules for them, to begin with. These were that each cab was to have a driver, that the driver should be able to stop the vehicle on demand, that the cabs could turn in a small space and that they should all be capable of climbing Savoy Hill, the steepest slope in London. To avoid unfair competition, the electric cabs would not be allowed to undercut the Hansoms, but would charge the same fares.

Londoners are well-known for the swiftness with which they coin colloquial expressions for any new or remarkable phenomenon in the capital and so it proved with Walter Bersey's electric cabs. They became known almost at once as 'Hummingbirds'. This was due in part to the garish yellow and black livery, which was a sharp contrast to the sober, all-black of the familiar Hansom Cab, but also because of the loud humming noise made by their electric motors.

The Hummingbirds had a top speed of 12mph, which was more than enough when moving through the congested streets of Victorian London. They were no faster or slower than the horse-drawn cabs. One distinctive feature, which did not meet with universal approval, was installation of electric light inside the passenger compartment. Electric headlamps on the front of the cabs was one thing, but harsh white light in the interior was felt by some to be a step too far. The dark interiors of Hansom cabs were seen as private and relaxing spaces and not everybody wanted to have bright lights illuminating them and making them visible to any passer-by in the street.

Because of the scarcity and unreliability of the public electricity supply in 1890s London, Bersey decided to erect his own power station, a colossally expensive undertaking. The cost of generating his own electricity, combined with various other problems with the fleet of taxi cabs, gradually worked to make the operation uneconomical. The cabs were very heavy, weighing two tons, which meant that the tyres wore out very quickly on the hard cobbled streets of the city. Because the technology was so new, the cabs also had a disconcerting habit of breaking down, something which seldom happened to the more dependable horse-drawn vehicles. Within two years, the Hummingbirds had vanished from the streets of London, an exciting experiment which was at least a century too early.

Of course, electric Hansom cabs appear in the novel *A Transatlantic Tunnel, Hurrah!* Here, they are gradually supplanting the horse-drawn version which is still, in this alternative 1973, popular. After hearing about the, 'new-fangled conversions that were slowly removing the presence of horses from central London', we are given a detailed account of the new, electric cabs:

> Here there was no proud, high-stepping equine frame between the shafts, but instead a squat engine of some sort whose black, metal, bricklike form rested upon three wheels. The single front wheel swivelled at a tug upon the reins bringing the hansom up smartly to the curb, while a tug on another rein stopped the power so it glided to a halt.

Once again, the strange, alternative universe which has been carefully crafted owes at least as much to historical research as it does to the vivid imagination of the author.

Electric cars flourished for a time in the United States, where they certainly gave vehicles powered by internal combustion engines a good run for their money. The world's first motor hearse, for instance, made its appearance in Buffalo, New York, in May 1900. It ran on a battery and there were no fewer than fourteen electric cars in the procession to the graveyard on what might be termed its maiden voyage.

Electric cars were popular in America as little city 'runabouts', similar in many ways to the Smart cars of our own time. They had low maximum speeds and limited range, but for city use, these were not major disadvantages. It was the discovery of huge new deposits of oil which brought about the decline of the electric car in the last few years before the First World War. Once cars with internal combustion engines were cheaper to run than battery-powered ones, then the writing was on the wall. More and more roads leading from city to city were now being well-surfaced with macadam and that in turn opened up the opportunity to use cars not merely for gadding about the city streets, but for long-distance journeys. This was where the limited range of cars which needed to have their batteries charged for hours after travelling just 30 or 40 miles began to have a serious disadvantage. With a car fuelled on petrol, it took only a few minutes to fill it up and be ready to go.

It is impossible to leave the topic of Victorian electrically-powered transport without mentioning one final mode of travel, which looks like a classic illustration for a steampunk story. Illustration 11 shows a most extraordinary vehicle, namely a seagoing electric tram. In 1883 a man called Magnus Volk opened an electric railway on the seafront at Brighton; a seaside resort in the south of England. This narrow-gauge railway is still in operation, which makes it the oldest operating electric railway in the world. Ten years later, Volk wanted to extend the railway line from Brighton to nearby Rottingdean, which was almost three miles away. The topography of the land, with undulating land and chalk cliffs, was not favourable terrain along which to lay railway tracks and so Volk hit upon the notion of setting the tracks along a much smoother and flatter part of the coast. He thought that the best idea would be to lay his tracks along the seabed, just off the coast, and then to run his trains through the sea itself. Not on a viaduct, but actually under the waves.

Of course, electricity and water are not a happy combination and so Magnus Volk attached his railway carriage to the top of 23ft-high stilts. The live cable which supplied power was suspended above the water and a pickup, rather like that used on trolley buses enabled the train to collect power. This may be seen in the illustration. Because at high tide this strangest of all trains was ploughing through water which was 15ft deep, lifebelts had to be provided for passengers; surely a first for a railway. Even more bizarre was that as the vessel was technically ocean-going, the law required that it be under the control of a qualified sea captain.

The Brighton and Rottingdean Seashore Electric Railway opened for business in November 1896 and was from the first tremendously popular. Posters advertised the service as a 'sea voyage on wheels' and for most passengers the trip through the water, 6d each way, was undertaken as a novelty rather than because they actually wished to visit Rottingdean. The world's only seagoing railway closed in 1901, when the local council began construction of a breakwater which would have required the tracks of the railway to be re-laid if it was to continue. There were no funds for such capital expenditure and so the service was simply discontinued. Pictures, like that in Illustration 11, are so much like a modern piece of steampunk art, that including some slight account of this means of travel was irresistible.

Volk's sea-going electric railway train was not, surprisingly, a one-off. The Victorians had a mania for innovation and this extended even to perfectly satisfactory structures and modes of travel which had served their purpose for millennia. Take bridges, the design of which had remained largely unaltered since the Middle Ages. They were essentially immobile and permanent structures. Several Victorian engineers devised what were known as 'rolling bridges', which were almost identical in appearance to Brighton's electric marine railway. The rolling bridge entailed building an underwater railway track and then placing on it a moving structure, just like Volk's at Brighton. This would then move back and forth across the river like a ferry on wheels. Very few were constructed, because of course unlike ordinary bridges of the kind with which we are familiar, they required constant attention and regular maintenance, and were very expensive to run. Like the ocean-going electric train, they were soon consigned to the dustbin of history.

One Victorian idea which very nearly came off and would have turned London into something which would not have been out of place in a science fiction story of the distant future, was a plan to cover part of the city in glass and so exclude the weather from the streets. This vision, which may be seen in Illustration 15, is the subject of the next and final chapter.

Chapter 9

The 11-Mile-Long Shopping Mall
that Never Was

We are used to finding strangely distorted and futuristic versions of the Victorian world when we read steampunk stories. That is indeed the whole point of the genre and if the nineteenth century depicted in novels such as *The Difference Engine* did not differ radically from the historical one with which we are all familiar, then such stories would not be steampunk at all. In this book we have looked at real Victorian inventions and seen how very nearly these transformed the world of that time into something resembling a steampunk fantasy; things such as the vacuum railway tubes, mechanical computers and steam cars which never quite took off in the real world which we know from history books. We have examined, too, plans which were drawn up in the nineteenth century and very nearly executed, which would at a stroke have turned the nineteenth-century world into something more like a steampunk novel. The world's first international airline, for instance, which it was hoped would enable a regular service of steam-powered airliners to travel between Britain and India. An airline in the 1840s operating a fleet of transcontinental steam planes was never likely to become a reality. Some of what sound to us now as pretty outlandish schemes though, ideas which were being floated at the height of Victorian prosperity, were completely practical and well within the scientific and technological capabilities of that time. One of these was a self-contained glass-covered city, with a carefully controlled pollution-free atmosphere, integral transport system, parks, houses, offices and of course shops.

The Victorians had a fascination for large indoor spaces with glazed roofs; places where one could forget the weather and enjoy a clement climate at any time of year. In the winter, when the outside world was cold, wet, muddy and smoky, it was pleasant to stroll in wide, public areas where there was no rain and everything was warm and dry. The first manifestations of this love of indoor public space, combined with shopping and social life, were to be found in the arcades which sprang up across Europe in the early nineteenth century. We sometimes imagine that shopping malls are a post-Second World

War, American phenomenon, but this is not so. It is true that the first actual climate-controlled shopping mall was opened in America in 1956, but the idea had a long history before then.

In the Middle East, covered bazaars had been operating from time immemorial in cities like Istanbul and Jerusalem. Perhaps inspired by the example of the Levant, a fashion arose in Europe in the late eighteenth century which caused that period to be known by some as *l'Ere des passages couverts* or the 'Arcade Era'. The craze for covered shopping streets began in France and spread to Italy, Russia and Britain. In London the Burlington Arcade opened in 1819. This consisted of seventy-two small shops in a wide passageway, with light entering through a glazed roof. Two hundred years later, Burlington Arcade is as popular as ever.

Shopping arcades where one would not be vulnerable to the vagaries of weather were one strand which led towards the idea of a covered city. Another was the enthusiasm for large greenhouses of the kind built in Kew Gardens in the 1840s. These were so constructed that fully-grown trees flourished in them and many people at a time could visit and admire the exotic plants which they contained. Grand houses had had orangeries for years without anybody seeing them as being of particular interest, but these public greenhouses seemed to capture the imagination of Victorian Londoners and it was not surprising that when a vastly bigger greenhouse was erected in the heart of the capital, it should prove an enormous success and an object of enduring fascination for millions of people.

In 1851, when the Great Exhibition, or the Great Exhibition of the Works of Industry of All Nations to give it its full and proper name, was being planned, there was a lot of debate over the most suitable structure to house the exhibition. Joseph Paxton, a gardener who had designed and overseen the construction of some of the largest greenhouses ever built, put forward a plan modelled on the huge greenhouse which he had created at the Duke of Devonshire's home at Chatsworth. The great advantage of Paxton's plan was that the building would be constructed from pre-fabricated sections of cast iron and sheets of glass which could be erected and taken down again in a matter of weeks. It would be over 1,800ft long, 120ft high and cover 22 acres. No lighting would be needed because the walls and roof would all be glass, hence the nickname which the great hall soon acquired; it became known universally as the Crystal Palace.

The Crystal Palace was built in Hyde Park and it enclosed a number of mature elm trees, which were left *in situ*. Over six million visitors made their way to the Crystal Palace in 1851 and although they gazed in wonder at the stuffed elephant from India and the nugget of gold from South Africa which weighed hundreds of pounds, for many people the most astonishing sight of all was the glass building housing the exhibition, which was three times the size of St Paul's Cathedral. Joseph Paxton had devised an ingenious way of ventilating his enormous glasshouse, by means of louvered openings which allowed stale air to escape. Even at the height of summer, the interior of the Crystal Palace was pleasant and fresh.

Visitors to the Great Exhibition wrote of the magic of the Crystal Palace and the sense of awe which it evoked. The chief wonder was the way that outdoor and indoor merged and became, in a sense, irrelevant. The towering trees of Hyde Park, around which the Crystal Palace had been built, were no longer something to be experienced in the open air, but had become, like the palms in Kew Gardens, more like house plants. One German visitor did his best to express this idea, when he wrote that, 'The total effect is magical, I had almost said intoxicating. The incessant and never-ending motley of forms and colours, the transparency on every side . . .' The Crystal Palace triggered a longing for similar projects, spaces where the public could mingle and shop, even perhaps live permanently, under cover and away from the elements.

So greatly did the Crystal Palace appeal to Victorian London, that when the Great Exhibition ended, nobody wished to see it pulled down and destroyed. Instead, it was carefully dismantled and moved to a site in south London, where it remained until it was destroyed by fire in 1936. What was the real attraction of the building? It was part greenhouse and part living space. Life in the streets of London at that time was often quite unpleasant. When it rained, the streets turned to mud. Throughout the winter, coal fires poured sulphurous and corrosive smoke into the air, sometimes resulting in smog (although the word itself was unknown at that time). At any rate, the air became sharp and unpleasant to breathe and the sooty smoke made everything grimy. Summer could mean stifling and uncomfortable heat, combined with unpleasant smells. On a boiling-hot August day, Victorian Londoners were not able strip to the waist, wear shorts or even roll up their sleeves, unless they belonged to the labouring class. Instead, they were obliged to wear as many layers of restricting clothes as in winter. Summer in the nineteenth-century British city was an exceedingly sweaty and uncomfortable affair. In the capital, smells from the inadequate sewage system were a perennial problem in warm weather. Sewers discharged into the Thames, and in central London the stench could be abominable. None of these unpleasant things were apparent when under glass in a place such as the Crystal Palace, where the air was cool and the smells and smoke of the city were kept out.

Apart from the bad air and muddy streets of London, it was an exceedingly congested and crowded city in the middle of the nineteenth century. There were only two ways of travelling about in the capital and that was on foot or by the use of horses, either riding them or in vehicles drawn by them. If one wished to get from the financial centre of the City of London, the area around the Bank of England and the Stock Exchange, to the West End, then it was either walking or a taking a horse-drawn bus or cab. The traffic at this time proceeded at less than walking pace and crossing central London was a slow and torturous affair. Then, as now, when nobody knew what to do about a problem, an enquiry seemed the most logical way to proceed. In 1855, a Parliamentary select committee was set up to look into various schemes which might improve matters. The Select Committee on Metropolitan Communications met in the first half of 1855 and published its report in October of that year, after having heard many witnesses suggest a number of schemes for relieving the chronic congestion of London's traffic. Some of the ideas were bizarre and others radical, but practical. Following the publication of the select committee's findings, the magazine *The Spectator* announced that, 'The Committee have evidently a leaning towards Sir Joseph's proposals . . .' What had Sir Joseph Paxton, architect of the Crystal Palace, suggested which had found such favour with the select committee?

When giving evidence to the select committee, Joseph Paxton had famously remarked that it took less time to travel from Brighton to London Bridge Station than it did to get from London Bridge to Paddington Station, just a few miles away. The problem was that all the great London railway stations such as Paddington, Euston, Kings Cross and Waterloo were not only some distance from each other, but also a stiff walk from the West End and central London. Luckily, he had a solution. How if all the major railway termini of London were linked together by a rapid transit system? This would mean not only that a traveller could easily get from one station to another, this new and speedy means of travel would also run into the heart of the West End, with a branch leading to Piccadilly Circus. *The Spectator* explained to its readers what Sir Joseph had in mind;

> Sir Joseph Paxton's plan for a grand girdle railway and boulevard under glass, with shops and houses attached, crossing the river three times, once a little way above Southwark Bridge, a second time between Waterloo and Hungerford Bridge, in connexion with

a branch from the New Cut to Regent's Circus, and again near Lambeth Palace. This vast scheme contemplates reaching all the railways coming into London; the connexion with the Shoreditch and London Bridge stations being effected by short branches. It possesses in its details, as will be found; on a reference to Sir Joseph's evidence, many features of remarkable novelty.

What was being proposed by Sir Joseph Paxton was not only a new way of transporting people quickly from one London railway station to another, but that this should also be incorporated into a new type of living space; a glass-covered city street which would stretch for 10 miles around central London.

Paxton's suggestion amounted in fact to a proposal not only for a transport network, but for an entirely novel way of living in a city. He wanted to build a new and improved version of his Crystal Palace, but this would not only be a location which one would visit to shop or see the sights. People would actually live in what he had provisionally named the Great Victorian Way. Shopping in the Burlington Arcade was a pleasant interlude, as was visiting the palm house at Kew Gardens. This new project though would be a cross between a shopping arcade and a greenhouse and it would contain not only shops, but also houses and an atmospheric railway, where the trains would not need to be drawn along by locomotives discharging steam, smoke and soot. The Great Victorian Way would be a miniature, self-contained city, where people could live, work, shop, visit their friends, walk in the park and travel around London; all without going outside into the open air.

It is hard now to understand just how revolutionary Joseph Paxton's idea was. Those living in London are today so used to jumping on a Tube train and being whisked speedily to any part of the city, that they cannot imagine how difficult it once was to get from place to place. One could either walk or travel by an omnibus or cab through streets so crowded that traffic proceeded at an agonizingly slow crawl. There simply was no means of moving about the largest city in the world faster than walking pace.

This is another example, as we saw with the internal combustion engine in an earlier chapter, when the way that history developed seems, in retrospect, inevitable. Residents of, and visitors to, London are so used to Tube trains that it never crosses their mind for a moment that any other way of getting around the city, other than diving beneath the earth, might have been considered in the past. When he was giving his evidence to the select committee which was examining the problem of transport in the capital though, Joseph Paxton expressly dismissed the idea of an underground railway, saying, 'People,

I find, will never go much above the ground, and they will never go under ground.' Those who advocated railways running beneath the ground had an uphill struggle to persuade people in authority that such a scheme would be practical or, if built, popular with passengers.

For Joseph Paxton and one or two other visionaries, the building of a new railway system which would actually run through central London was not just a way of solving a difficulty in transporting people from one place to another. It was rather an opportunity to change the whole way that city life was viewed. City dwellers are used to being at the mercy of the vagaries of the weather. They expect to get wet when it is raining and to swelter in the streets during a summer heatwave. True, when they get home they can be warm and dry in the winter and even, if they install air-conditioning or heat pumps, fresh and cool in the hottest summer. Despite all the technological advances made since 1855, we still draw a sharp distinction between conditions inside our homes and those in the street outside. The one we can, within certain limits, control. The other is wholly unpredictable and we just have to put up with it as best we can. If it snows when we are outside our homes, we get cold and wet. If it is a hot, sticky day, then we will sweat in the heat.

The Great Victorian Way was essentially an extended version of the central part of the Crystal Palace; repeated over and over again for a distance of 11 miles. A broad street, 72ft wide, would run along the centre of a glazed arcade which soared over 100ft overhead. Shops and houses would line this street. No fewer than eight atmospheric railways would also run alongside the houses. Four of these railways would be for a slow service, stopping at frequent intervals; the others would be expresses which would take people straight from one railway terminus to another with no delays. Sir Joseph had carefully worked out the cost of such a system and it came to £34,000,000. He believed that the initial capital cost, borrowed from the Treasury, could be recovered from the rent paid for houses and shops. His calculations suggested that around 105,000 passengers a day would use the railway and this too would produce a healthy revenue. A cross-section of the Great Victorian Way, as envisaged by Paxton, may be seen in Illustration 11.

There was no doubt that this idea, of an enclosed and self-contained living space in which the climate would be completely under control, appealed to the members of the select committee. The Great Victorian Way would run in a vast loop from the Bank of England, south across the Thames and then sweeping west towards Waterloo Station. From there, it would cross back over the Thames near the Houses of Parliament, then through Sloane Square and north into Kensington Gardens. It would run to Paddington and swing back east, linking up to Euston and King's Cross stations, before returning to the Bank of England. This would be a distance of about 10 miles. A spur would run across the Thames from Waterloo Station to Piccadilly Circus and other spurs were envisaged, which might be built to connect to other stations in the future.

Everything had been thought of by Joseph Paxton. The three bridges across the Thames would also have houses and shops on them, making them the first inhabited bridges in the country since the houses on the old London Bridge had been demolished in the middle of the eighteenth century. Even relaxation and leisure had been considered. The part of the Great Victorian Way which ran through Kensington Gardens would have no buildings. Instead, it would enclose an area of open parkland where people could walk when they wished for some exercise. As the great glass girdle was described, it would be a wholly self-contained little world, where residents could live without ever needing to go to any other part of London, unless they particularly wished to do so.

There is definitely a touch of science fiction about the plans for this enormous, glass-covered living space. Inhabited space stations have been discussed, which would be in the shape of great wheels and contain even parks and trees. The images of such fantastic schemes resemble the Great Victorian Way, glass-covered circles which contain all that is necessary to life.

No idea, however revolutionary, occurs in a vacuum and Joseph Paxton's plan for a huge, covered boulevard was not wholly original. Others had in the past put forward such plans and even as Sir Joseph was urging the adoption of his own project, others in the country were also proposing similar schemes for London. All were perhaps inspired by various suggestions made in France for the building of a glazed roof over boulevards and parks in Paris. None of these had actually come to anything, but the idea had, by 1855, certainly been floating about for years. William Mosely came up with something very similar to Paxton's Great Victorian Way, rather cheekily making an explicit reference to the Crystal Palace in his proposed Crystal Way. Like Sir Joseph, Mosely envisaged a glass-covered street along which vehicles would run. A railway would run beneath this street, constructed as a cut-and-cover tunnel, similar to those used in the first underground railways which would be built in London from the 1860s onwards.

Frederick Gyle, who was in 1855 the director of London's Royal Italian Opera in Covent Garden, had already caused something similar to the Crystal Palace to be built alongside the opera house. The Floral Hall, really a gigantic greenhouse of iron and glass, still stands next to what is today the

Royal Opera House. Ten years before all the talk about the Great Victorian Way and Crystal Way, Gyle had written about the possibility of a large, glazed arcade in central London. This would enclose cafes, exhibition halls, shops and various other public buildings. Gyle had in mind something far more like a modern shopping mall, which would be entirely pedestrianized. There was no provision in his arcade for carriages or railways.

Prince Albert, the prime mover behind the Great Exhibition, was enthusiastic about the Great Victorian Way and believed that it would revitalize and modernize London. The involvement of the Prince Consort meant in turn that this scheme had the tacit approval of the Queen herself. It had perhaps been a sharp move to gain royal approval which had prompted Sir Joseph Paxton to name his project after her majesty. Now that the select committee appeared to be backing the idea as well, it looked as though nothing would stop it. The sticking point, which ultimately proved fatal to the whole enterprise, was the initial cost. The British government in the 1850s was busily engaged in empire-building abroad, which cost an awful lot of money. The spirit of the times was very much in favour of private enterprise and people helping themselves, rather than relying upon central government to foot the bill. If Paxton had been able to raise the necessary capital, then it is possible that the government might have smoothed the way a little, by clearing the land needed for the Great Victorian Way and generally lending a hand here and there. What they baulked at was footing the whole of the cost and then waiting many years for the expected revenue from the railways, rents from buildings and so on to repay the loan with interest.

It is not usually in the nature of British governments to make rapid and irrevocable decisions and nobody in government wished to reject outright an idea which apparently had the backing of Queen Victoria and her husband. Instead, it was, as we say today, put on the back burner, while it was seen whether or not private investors might be inclined to put forward some of the money. £34,000,000 really was an enormous sum of money in those days and there were probably more important things on which it might be spent. In fact, the money was in the end spent and left a legacy in London from which the inhabitants still benefit every day, although they never see the superstructure upon which the money was expended. It is a scheme considerably less attractive visually than a glass-covered boulevard and railway would have been.

In 1854, the year before the Select Committee on Metropolitan Communications met, there was a serious outbreak of cholera in London. This centred around Broad Street in the West End. Over 600 people living in the area of this street died in the epidemic. At that time, it was widely believed by the medical profession that illness was caused by 'miasma'; that is to say, infections were carried on bad air. The smell of rotting organic matter or excrement was thought to be the agent which spread disease. A rival theory was becoming popular though, which was that 'germs' were really what allowed infections to be transmitted from one person to another. It was increasingly being thought that an illness such as cholera was actually carried in contaminated water or by bodily contact, rather than on the air by 'miasmas'.

During the London cholera epidemic of 1854, a doctor called John Snow decided that the disease was centred around a water pump in the street. Because few working-class homes at that time had plumbing, water for drinking and washing had to be fetched regularly from a standpipe in the street. If Dr Snow was right, then contaminated water from this source could be the root cause of the deaths which were occurring with increasing rapidity in the neighbourhood. There were no flushing lavatories in the district and chamber pots were emptied into a cess pit which was only a few feet from the pump. After putting his theory to the authorities responsible for the pump, the handle was removed and the epidemic halted in its tracks, confirming almost overnight the new germ theory of infection.

The problem in London of the 1850s was that human waste was disposed of by either being buried in the ground or dumped in the River Thames, which was in consequence an open sewer. This was common enough at that time. There is an apocryphal, but altogether plausible, story that when Queen Victoria visited Trinity College at Cambridge University, she observed the state of the river Cam and innocently asked the warden, 'What are all those pieces of paper floating past?' Thinking on his feet, the man replied, 'They, ma'am, are notices that bathing is forbidden.' In London, the faeces-laden water from the Thames was drawn out by water companies and piped into people's homes.

In 1858 a hot summer resulted in London's 'Great Stink', when the smell from the sewage clogging the Thames became so bad that even Parliament was forced to take notice and do something about the situation. The stench was so awful in the House of Commons, which stands by the river, that it was seriously suggested that the legislature and government might have to be removed to Oxford or St Albans. Public money was now made available for the construction of an effective system of sewers for London. The expenditure on new sewers meant that there was certainly nothing left in the kitty to pay for the building of either the Great Victorian Way or the Crystal Way.

The problem of traffic congestion in London was about to be solved anyway, without its costing the government a penny. For years, money had been raised for an underground railway, which began to be built in 1860. When it opened a few years later, it was so enormously popular that it was at once seen that this was the answer to the problem of the capital's congested streets. Despite Joseph Paxton's claims that people would not go underground, the public voted with their feet and London's Tube network was the result. As a matter of fact, a pale shadow of the Great Victorian Way can still be discerned to this day in London. The Circle Line follows closely the route for Paxton's mooted atmospheric railways and glazed boulevard.

Although neither the Great Victorian Way or the Crystal Way were ever built, the influence of such ideas lingers on, if on a much smaller scale than originally envisaged. The great railway stations of Victorian Britain are nothing more than enormous glazed arcades. They contain shops, bars and hotels, were designed from the beginning to be impervious to bad weather and functioned as meeting places and shopping malls. If one wishes to see what the Great Victorian Way might have looked like, had it ever been built, one need only stand near the shops in London's King's Cross Station and look upwards. We take those great glass roofs for granted and hardly even give them more than a passing glance, but this is exactly what Paxton's 11-mile-long street might have looked like.

The idea of the self-contained, weatherproof public space continued to intrigue the Victorians and they toyed with various different versions of it in the years following the abandonment of the grand proposals for streets and parks covered with glass. Covered markets were one way of experimenting with what Paxton, Mosely and Gyle had been thinking of. London's Leadenhall Market, for example, is like a vastly scaled-down miniature version of the Great Victorian Way. These covered streets were built in the 1880s, on the site of what was once Roman London's basilica. This is a shopping arcade unaffected by the weather. Just like the railway terminuses of the time, it gives us an idea of how life in Victorian cities might have become, had things turned out just a little differently.

The thought of climate-controlled Victorian cities, whose inhabitants never needed to leave their comfortable environment and would never be inconvenienced by a shower of rain or a smoky chimney, is an enchanting one and has a definite flavour of steampunk about it. It is just one more aspect of the nineteenth-century world which, along with the widespread use of steam-powered cars and the adoption of modulated light beams to carry telephone conversations, did not come to pass. It is not hard though to see how Paxton's grand idea could have taken off and spread. The Victorians were great ones for subduing nature and bending it to their will. There was no nonsense with them about respecting the natural environment or anything of that sort. A city immune to the weather, without muddy streets or stinking rivers would have been a very tempting prospect.

We might not realize it, but this vision was at least partially brought about. It was perfectly possible for somebody in Victorian England to travel about all over the place without actually coming out into the open air. One could begin in one of the railway stations in the north of England, Manchester say. These large Victorian stations were, as has already been remarked, essentially covered arcades full of shops, cafes and bars. A high glass roof protected people in the station from the weather outside. It would be possible to board a train at Manchester, travel to London and then get off the train, again under cover, and find one's self in a similar glass-covered arcade to the one which one had left, hours earlier and hundreds of miles away. After that, one could enter the Underground and travel by Tube train to anywhere in the capital, still without encountering any weather. We take sheltered travel so much for granted now, that most of us don't even realize what a novel idea it was 150 years ago.

It is almost impossible to imagine what London and perhaps the other large cities of Victorian Britain might have been like if schemes along the lines of the Great Victorian Way had left the drawing-board and been realized. Even today, 170 years later, the thought of walking for miles around a city without having to worry about rain or the cold is an enticing one.

Endword

Since the earliest appearance of proto-steampunk novels such as *Queen Victoria's Bomb*, *Pavane* and *A Transatlantic Tunnel, Hurrah!*, many readers have marvelled at the imaginative power of authors able to dream up such things as a continent-wide network of semaphore towers, which serves the same purpose as telephones and radio in our own world. The idea of a rapid-transit system which entails sending passengers through vacuum tubes at high speed also seems far-fetched and likely to have been conjured up by a fiction writer for our amusement. As we have seen, both these things actually existed in the real world of the nineteenth century.

The world created by the authors of steampunk novels and short stories strikes us as outlandish and fantastic, simply because we do not know enough about the history of the real world in which we live. It is hard to find a single invention mentioned in steampunk literature which did not have its counterpart in the age of the Victorians. One might almost say that to be a good author of steampunk, a vital prerequisite is to be well versed in the history of nineteenth-century technology. Combined with a knowledge of history, the aspiring writer of steampunk fiction would also do well to familiarize him or herself with the popular literature of the era. The most successful modern steampunk draws indiscriminately from both sources. When done skilfully, this can result in something so realistic that readers will be unable to unpick the fantasy from the solid, historical truth. We have seen how this worked so well with the Victorian robot Boilerplate, that a third of those encountering Boilerplate on the Internet believe that it was a genuine invention. Not for nothing has Boilerplate been called a 'gateway drug to steampunk'. Seldom has fact and fantasy been more effectively combined.

Those reading the first acknowledged steampunk novel, *The Difference Engine*, will not fully appreciate it unless they have a comprehensive grasp of the historical events upon which it is based. Without knowing about inventor Goldsworthy Gurney, for instance, references to 'gurneys' will mean nothing. The same is true of all steampunk. This necessity to have a good understanding of history, fully to understand steampunk narratives, has been a feature of the genre from its earliest days. As far back as *Bring the Jubilee*, published in 1953, an eye for obscure history was required to catch all the nuances of the tale. We are told in *Bring the Jubilee* that the Mormons live in a state called Deseret. This was in fact the name of a very short-lived provisional state, declared by the Mormons in 1849 and soon put an end to by the government in Washington. In this alternative world, Deseret still exists. For any reader lacking this historical knowledge though, the allusion would be quite lost.

In this book we have merely skimmed over the surface of some aspects of Victorian life and shown how they have become incorporated into modern books. We have also come across remarkable inventions and encountered technology which was decades or centuries ahead of its time. These two strands, actual inventions and incidents from the past, together with images and ideas from popular nineteenth-century literature, were combined in the earliest proto-steampunk novels at which we have looked in this book. These novels in turn laid the foundation for the steampunk movement which emerged in the late 1980s and early 1990s and has since then gone from strength to strength.

When next we come across what appears at first glance to be an extravagant piece of fantasy in the steampunk genre, it might be worth undertaking a little research. We shall invariably find that rather than being just the free creation of some author's imaginative faculty, whatever we are examining will turn out to have its roots in the real world. From steam-powered aeroplanes, cars or computers to robots like Boilerplate, the origin of almost every feature of modern steampunk is to be found in Victorian books, magazines and newspapers. Steampunk, although generally described as a genre of science fiction, has in fact far more in common with science fact than most aficionados ever realize.

Bibliography

Amis, Kingsley, *The Alteration*, Johnathon Cape, 1976.

Bellos, Alex, *Alex's Adventures in Numberland*, Bloomsbury Publishing, 2010.

Beyer, Rick, *The Greatest Stories Never Told*, HarperResource, 2003.

Camp, L. Sprague de, *Lest Darkness Fall*, Henry Holt, 1941.

Challoner, Jack, *1001 Inventions that Changed the World*, Cassell Illustrated, 2009.

Clark, Ronald, *Queen Victoria's Bomb*, Johnathon Cape, 1967.

Clayton, Antony, *Subterranean City: Beneath the Streets of London*, Historical Publications, 2000.

Clayton, Antony, *The Folklore of London*, Historical Publications, 2008.

Deary, Terry, *Dangerous Days on the Victorian Railways*, Weidenfeld & Nicolson, 2014.

Dickens, Charles, *Hard Times – For These Times*, Bradbury & Evans, 1854.

Gibson, William, & Sterling, Bruce, *The Difference Engine*, Victor Gollancz, 1990.

Grimley, Gordon, *The Origins of Everything*, Odyssey Press, 1973.

Harris, Melvin, *Book of Firsts*, Michael O'Mara Books, 1994.

Harrison, Harry, *A Transatlantic Tunnel, Hurrah!* Faber & Faber, 1972.

Hart-Davis, Adam, *Eurekaaargh!*, Michael O'Mara Books, 1999.

Homer, Trevor, *The Book of Origins*, Piatkus Books, 2006.

Horton, Edward, *The Age of the Airship*, Sidgwick & Jackson, 1973.

Moorcock, Michael, *The Warlord of the Air*, New English Library, 1971.

Moore, Ward, *Bring the Jubilee*, Ballantine Books, 1953.

Pratchett, Terry, *Going Postal*, Doubleday, 2004.

Pullman, Philip, *Northern Lights*, Scholastic Point, 1995.

Reeve, Philip, *Mortal Engines*, Scholastic, 2001.

Roberts, Keith, *Pavane*, Rupert Hart-Davis, 1968.

Robertson, Patrick, *The Shell Book of Firsts*, Ebury Press, 1974.

Slee, Christopher, *The Guinness Book of Lasts*, Guinness Publishing, 1994.

Standage, Tom, *The Victorian Internet*, Weidenfeld & Nicolson, 1998.

Sweet, Matthew, *Inventing the Victorians*, Faber & Faber, 2001.

Taylor, John W. R, Taylor, Michael J. H., & Mondey, David, *The Guinness Book of Air Facts and Feats*, Guinness Superlatives, 1970.

Tibballs, Geoff, *The Guinness Book of Oddities*, Guinness Publishing, 1995.

Trench, Richard, & Hillman, Ellis, *London Under London: A Subterranean Guide*, Book Club Associates, 1985.

Uhlig, Robert (ed.), *James Dyson's History of Great Inventions*, Constable & Robinson, 2001.

Verne, Jules, *From the Earth to the Moon*, Pierre-Jules Hetzel, 1865.

Verne, Jules, *Paris in the Twentieth Century*, Del Ray, 1997.

Webb, Simon, *Commuters: The History of a British Way of Life*, Pen & Sword, 2015.

Webb, Simon, *The Analogue Revolution*, Pen & Sword, 2018.

Weber, David, *Off Armageddon Reef*, Doherty Associates, 2007.

Index